More Praise for *The Amazon from Source to Sea*

"This is a gripping an elegantly written saga, lush with color photographs, that effectively puts a paddle in our hands and takes us down the world's mightiest river. And thanks to West Hansen's humble account, we need to rethink both the Amazon's source and demise of the river in postmodern times."

— Jon Waterman, author of *Atlas of the National Parks and Arctic Crossing*

"West Hansen's sprawling, fast-paced narrative is by turns harrowing and humorous, but always refreshingly honest. West's audacious vision and relentless drive reminded me of some obsessive film director—say, Werner Herzog making *Fitzcarraldo* or Francis Ford Coppola working on *Apocalypse Now*—turning an idea into an awe-inspiring reality in the face of daunting physical and psychological adversities. The astonishing complexity of Hansen's undertaking—terrain, politics, people, paddle craft—and West's inviting prose immediately marks *The Amazon from Source to Sea* as a landmark adventure book. I highly recommend it to armchair adventurers and expedition paddlers alike."

— Ed Gillet, only person to have solo kayaked from California to Hawaii and subject of the book *The Pacific Alone* by Dave Shively

"Fifty-year-old canoe racer-turned-explorer West Hansen is planning to kayak the Amazon River when he learns that everyone—cartographers, adventurers, even his own sponsor, National Geographic Society—has misplaced the source of the world's greatest river. One of exploration's great prizes is suddenly back up for grabs, and to claim Hansen has to shepherd a team of irascible Texans and international whitewater stars some 4,200 miles, from the crest of the Peruvian Andes to the Atlantic Ocean. The journey brings him face-to-face with a controversy as old as Livingstone's quest for the source of the Nile, in addition to the usual obstacles. With great humor and insight, Hansen details a wild ride full of personality conflicts, extortion, Machiavellian subterfuge, pirates, drug lords, uncharted whitewater, massive thunderstorms, injuries, illness, fatigue, tropical heat, blizzards, altitude sickness, jungle drunks, bales of marijuana, substandard scotch, bureaucratic labyrinths, loneliness, colossal tides and the unstoppable force of the largest and longest river on the planet."

— Jeff Moag, Freelance Writer and Editor, former editor of *Canoe and Kayak Magazine*

The Amazon
From Source to Sea

The AMAZON
from Source to Sea

The Farthest Journey Down
the World's Longest River

West Hansen

Worldwide Waterways, Inc.

Tom,

Thanks so much for joining our team for the Northwest Passage. Being an Arctic Cowboy may not get you free drinks @ the local watering hole; however, I sure appreciate all you're doing for the team. Please enjoy reading about our little Amazon float trip.

West
April 15, 2022

© 2019 West Hansen
All rights reserved. No part of this publication may be reproduced, stored in a retrieval system, or transmitted, in any form or by any means, electronic, mechanical photocopying, recording, or otherwise, without the prior written permission of the publisher.

Published in the United States by
Worldwide Waterways, Inc.
PO Box 1718
Groves, TX 77619
www.westhansen.com

Cover Design and Illustrations by Doug Brown
Interior Formatting by Danielle H. Acee

Library of Congress Cataloging-in-Publication Data
Hansen, West

The Amazon From the Source to the Sea: The Farthest Journey Down the World's Longest River

p.   cm.
ISBN   978-0-578-50973-0 (hardcover)

*In memory of my friends, Bill Grimes of Iquitos, Peru,
Patrick Falterman of Texas and Juanito De Ugarte of Cusco, Peru.*

Thanks to…

Manuscript peer reviewers, Barbara Edington, Celeste Richardson, Chris Kuykendall, Darcy Gaechter, John Erskine, Pam LeBlanc and Tim Curry for making some very valuable suggestions.

A special thanks to my editor, Jeff Moag, who helped me carve down my nightly journal notes, typed on the banks of the Amazon River, into a readable accounting of the expedition.

My sister, Suzie Pennington, cousins Jennifer and Jamie Aucoin and father, Ed Hansen for helping with my fundraiser. My mother, Ann Hansen, for doing super mom stuff. My wife, Lizet and daughter, Isabella, for tolerating my eccentricities.

Debbie Richardson, Sheila Reiter, Ginsie Stauss, my niece Monica Pennington and many more for in-kind, financial and on-the-ground support. All the people who watched us and followed the expedition on the internet. The canoe racing community of Texas and Missouri rallied behind me for this expedition and I really appreciate their support.

The Andean support crew: David Kelly, Jennifer Maxfield-Boucher, Piotr Chmielinski, Alex Chmielinski, Lucho Hurtado, Lizet Alaniz, Isabella Hansen, Jeff Wueste, John Maika and Jimmy Harvey. Most especially, David Kelly for helping me keep my cool and for dealing with aggressive journalists and expedition politics. David put up with my frequent Hulk moments and helped bring me back down to a manageable level.

The Tigers: Juanito De Ugarte, Rafa Ortiz, Tino Specht, Simón Yerovi, Daniel Rondón, Santiago Ibañez and Chava Loaya for keeping me alive on the Mantaro River.

Friends, César Peña, Jason Jones and Erich Schlegel for supporting the team on the Ucayali and Marañon Rivers and for such great photos.

Rocky Contos for sharing with me the information he found about the sources of the Amazon River.

Pete Binion and John Maika for joining us on the Ene, Tambo and Ucayali Rivers.

Yonel Guzman of Peru Maderas in Orellana, Peru.

Mauro and his staff of the Hotel Ches Les Rois in Manaus.

The Peruvian Coast Guard for keeping us safe in unsafe areas, especially Captain Alberto Alcalá, Captain Francisco Bolanos, Captain Bustamante, Lieutenant Carlos Avila and Ensign Emilio Huaco of the Peruvian Coast Guard. In Brazil, Commander Robson Neves Fernandes, Commander Paulo César Mochado, Lieutenant Julio Leite, Sergeant Vladimir, Captain José De Fatima Oliveira De Andrade, Sub-Lieutenant Dionaldo Reboucas Dos Reis and Lieutenant Adilson Da Silva Dias of the Brazilian Coast Guard.

Igor Vianna for his support in Belem.

Jeff Wueste and Ian Rolls for being such great friends and partners.

Table of Contents

Preface 1
Part One: The Mountains 5
   Chapter One: Fits and Starts 7
   Chapter Two: How, When and Why I Ended Up on the Amazon River 24
      The Mother of All Monkey Wrenches 27
      Sowing the Seeds 32
   Chapter Three: Back to the Very First Day of the Expedition 37
   Chapter Four: The Wild, the Innocent and the Peruvian Shuffle 46
   Chapter Five: The Mantaro Pushes Back 54
      The Tino Show 64
   Chapter Six: Mayhem and Mary Jane 69
   Chapter Seven: The Rocky Road Less Traveled 77
   Chapter Eight: The Vegan Adventure Tuber 85
   Chapter Nine: The Big Bend 89
   Chapter Ten: The Water Cannon 100
      Dynamite Canyon 105
   Chapter Eleven: The Middle Canyons 111
   Chapter Twelve: Thin Air and Dirty Water 127
      Drag and Drop 135
   Chapter Thirteen: Getting There is Half the Fun 141
      The Lower Canyons 145

Part Two: The Jungle 159
   Chapter Fourteen: Out of the Frying Pan 161

Chapter Fifteen: Ayacucho 167
    Thanks for Flying Fuego Airlines 171
    The Lucky Chicken 176
    Fredy's Resort 179

Chapter Sixteen: Shotguns on the Ene 182
    Peeled Faces Among the Unconquered 187

Chapter Seventeen: Puerto Prado 192
    Badges? We Don't Need No Stinking Badges! 197

Chapter Eighteen: Atalaya 207
    That Night in Atalaya 209

Chapter Nineteen: Death and Dismemberment on the Ucayali 213
    The Devil's Bend 216
    Galilea 219
    The Calvary 223

Chapter Twenty: The Pucallpa Accord 229
    The Honeymoon is Officially Over 236

Chapter Twenty-One: Contamana 241
    Through the Land of the Isconahuas 243
    Se Breve, Por Favor 246
    The Cat is Out of the Bag 249

Chapter Twenty-Two: Nothing but the Best 251
    La Gringa 257
    The Maranon 263

Chapter Twenty-Three: Parting Ways 271
    Iquitos 273
    Rest, Relaxations and Red Tape 279

Part Three: The Inland Sea 285
    Chapter Twenty-Four: Storm Alley 287

  Lepers, Revolutionaries and Rabbis 290

Chapter Twenty-Five: You, Again? 295
  The Land of Burning Embers 300
  Horizon Lines on the Solomoes 303

Chapter Twenty-Six: The Routine 309
  Tefe 314
  Shooting the Amazon 319

Chapter Twenty-Seven: Kindness and Comfort in Camara 325
  Manaus 329
  Lapping up Luxury 333

Chapter Twenty-Eight: Monsters on the Amazon 339
  The River Amazon 340
  The Obidos Narrows 345
  Juruti 347
  On the Island of Swimsuit Models 350

Chapter Twenty-Nine: The Riviera of the Amazon 353
  South of Dixie 357
  Lawyers, Guns and Money 363
  Catch and Kill 366
  An Escape from Paradise 368

Chapter Thirty: Big Water 371
  The Lower Amazon Grind 376
  Through the Heart of the Amazon 380
  The Rio Pará 385

Chapter Thirty-One: Tidewater 390
  The Incredible Heaviness of Being 395

Chapter Thirty-Two: The Maw 406

Epilogue 413

Preface

"If you march your Winter Journeys you will have your reward, so long as all you want is a penguin's egg."
            Apsley Cherry-Garrard, *The Worst Journey in the World*

A river is greater than its name.

All rivers and streams are composed of uncountable droplets of water flowing from many places to combine into one final stream to meet the ocean together. The most distant source is that particular place farthest from the ocean from which a drop of water must travel, regardless of whether that drop comes from a glacier, storm cloud, dewdrop, snowbank, swamp or pond and regardless of the different names assigned to it, or its route, or what hindrances it must navigate through, under, over or around to complete the journey.

It's that simple.

I did not paddle from the most distant source of the Amazon River to the Atlantic Ocean. Let's get that out of the way. I'm not the first Amazon River source-to-sea paddler to attempt this goal, and I'm honored to be amongst notable explorers to make the claim, only to learn differently after our effort when science and exploration prove otherwise. Almost two years after the expedition I learned that the most distant source of the Amazon is in fact a little stream adjacent to the one from which we launched our expedition, and it's about two and a half miles farther from the Atlantic. I did paddle farther down the Amazon River than anyone else in recorded history and was the first to paddle the entire Amazon from the Mantaro River watershed, which has recently been established as the most distant source in the entire Amazon watershed.

The goal of the explorer is less about the fleeting moment when they summit the peak, hit the shore or cross the finish line, than it is the journey. Never more is the adage that happiness is the path to happiness more applicable than on an expedition. An explorer's life is one of shifting desires, either for the warmth and comfort of home when in the wilderness and of the campfire, the close calls, the fatigue or the stars when enduring the often-frustrating comforts of a civil, modern existence.

Momentary homes on the trail or river, in the middle of what many naively consider "nowhere," carve deep scriptures into explorers' lives, making our lives and deaths more worthwhile.

All expeditions experience internal tension, personal conflict and political machinations. Most of these dramas play out far from the action, yet they have a profound impact on those sweating under the load. In these pages, I've tried to relate what we endured, failed and overcame on this expedition, as individuals and as a team. The most difficult part in telling this story has been portraying, as accurately and objectively as possible, the contentious aspects of our shared endeavor. As Jennifer Kingsley noted, quite truthfully, in her memoir *Paddlenorth*, "Everyone has a different reason for committing a journey to paper, and no one has the same memory." This book is my memory, augmented by my contemporaneous notes, hundreds of hours of video and the recollections of others who were there. With the help of my editor Jeff Moag and eight reviewers who provided frank and often pointed feedback, I strived to address such conflict only when it affected the expedition and to do so with fairness and honesty. Still, it's been difficult to write about problems I encountered with those I care about and even those I don't care so much about. After the fact, I felt much better upon reading Victoria Jason's *Kabloona in the Yellow Kayak* and Don Starkell's *Paddle to the Arctic*, with their competing accounts of their shared expedition and the rift that grew between them.

These words have been translated from notes I made each evening, knowing that the focus of such an expedition is so intense that some details would be quickly forgotten and others skewed with the passage of time. I've been surprised at how imprecise my memory has been about particular events when I've gone back to my expedition journal to verify the details.

I'm thankful to all those who contributed funds, gear, and support of every kind. Even those who appeared to present challenges to overcome contributed something along the way. It's pretty rare that any one person can accomplish such a journey without assistance, and I had a lot of it. Even the loneliest arctic explorers had someone watching their back at some point, if not along the way. My life was saved many times on the Mantaro by the Tiger team and our whitewater leader, Juanito De Ugarte. I'm especially appreciative of my sister, Barbara Edington, who took on the role of Expedition Manager, bore the brunt of my bad moods, picked me up when I was down and told me when I was being a jerk.

Our team paddled the entire length of the Amazon River, which is about the same distance from Austin, Texas to Tuktoyaktuk, Canada above the Arctic Circle, from Minneapolis, Minnesota to Paris, France, from Moscow to Bangkok, from London to Nigeria and from Florida to Morocco. The journey took 111 days (90 days of actual paddling on the river, with 21 days for rest and logistics), and each bit of forward progress was made of our own human power, with raft and kayak paddles. We had bank support for about 100 miles of the 500-mile Mantaro River and boat support for about 1,500 miles of the 3,800-mile flatwater section. Otherwise we were on our own.

Though my life has been somewhat of a tinderbox, this particular spark was lit after reading *Running the Amazon*, in which author Joe Kane recounts the journey he took with Piotr Chmielinski and others down the Amazon River. Before going beyond this preface, I strongly urge you to read Kane's fascinating book to gain some history and perhaps to put into perspective my mindset as I entered into this adventure, the hydrological issues, politics and personalities that often take center stage in the story you're about to read. If my story ignites even the slightest spark in one person, I'll consider all of the coffee, scotch and anguish I filtered through my body during the writing process well worth it. I owe a great deal of thanks to Joe, Piotr, Zbigniew and Kate for sharing their story, and appreciate their contribution to the world of Amazon exploration.

—West Hansen, 2019

Part One
The Mountains

Chapter One
Fits and Starts

August 18, 2012
Day 2, Mile 3.61

"I would rather be ashes than dust! I would rather that my spark should burn out in a brilliant blaze than it should be stifled by dry-rot. I would rather be a superb meteor, every atom of me in magnificent glow, than a sleepy and permanent planet. The function of man is to live, not to exist. I shall not waste my days trying to prolong them. I shall use my time."

Jack London

Snow storm on the Amazon. Photo by Erich Schlegel

Dawn came quickly in the highlands of central Peru, filling an overwhelming silence with clean bright light. Jagged snowcapped peaks loomed on three sides of the plate-glass Lago Acucocha, beyond which the altiplano extended forever. Whatever clouds separated us from the heavens the night before had blown westward, to the Pacific Ocean, wiping clean the celestial palate. Underfoot, frosted ichu grass crunched loudly as I carried a steaming cup of coffee to the van and tapped on the window.

Jeff Wueste pulled the breathing mask from his face and took a cautious sip. An inadvertent cough would send the precious elixir flying. He'd drained our last two oxygen canisters overnight, sitting upright in the van, wrapped in a sleeping bag.

"Let's get moving," he wheezed.

The day's goal was to kayak an 18-mile stretch of the Río Gashán, which dripped out of the lake and across the high lonesome plain, destined to become the largest and longest river on the planet. Before the Amazon River spilled into the Atlantic Ocean some 4,200 miles east and 14,800 feet in elevation below our little highland encampment, it would plunge through whitewater canyons, across lawless jungle and teeming cities before slamming violently into the formidable sea. The day before (the official start of the expedition) we paddled across Acucocha to pinpoint the trickle, marking the newly discovered source of the world's mightiest river. This pinpoint was discovered just four months earlier. We set out to run the Amazon for the first time from its most distant source to the sea, but all we had to do today was paddle a few easy miles. I clapped Jeff on the shoulder and went off to pack my kayak.

I chose to launch our expedition in early August to take advantage of the dry season, when the normally flooded Amazon watershed carried far more manageable flow, with fewer monsoons and much better options for campsites along the rivers.

Our alpine camp bustled with 14 people, including the ground team and two distinct paddling tribes. John, Jeff and I are marathon canoe racers from Texas, none of us less than 50 years old. The four "Tigers," as we called them, were part of an international brotherhood of whitewater boaters whose job was to get us safely down the Mantaro River, a 500-mile whitewater gauntlet that had only been run once before. Since there was no hard whitewater on the docket today, we Texans would paddle three of the four kayaks. Rafa Ortiz, a gregarious young man from Mexico City who'd recently gained fame by paddling off a 189-foot waterfall on purpose, had drawn the short straw and would babysit us on the flatwater.

Jeff, John, Rafa and West portage from Lago Acucocha into the Río Gashán.
Photo by Erich Schlegel

Sealed up in my rubber-gasketed dry-top, I sat in my kayak oozing sweat as the Tigers helped Jeff and John into their boats. Jeff looked quite puny and John looked just plain old, unsure of what to make of the tight plastic kayak. He'd logged thousands of miles in narrow racing canoes, but that's as different from whitewater kayaking as two types of paddling can get. Still, if John and Jeff were to paddle the entire Amazon River, they had to start here.

My own whitewater skills were outdated and underused, although I started as a whitewater kayaker in the 1980s. Jeff and John's experience with rapids was little and none, respectively. Still, once past the Lago Acucocha spillway 200 yards ahead, we'd have nothing but shallow, bottom-dragging streams for the next couple of days.

The sun shone brightly and team members strolled alongside us, chatting and snapping pictures. My old friend and photographer Erich Schlegel, on assignment from our chief sponsor National Geographic, looked like a young Dennis Hopper, a la *Apocalypse Now*, in his khaki photographer's vest and wide-brimmed hat. With him walked Mauro, an ancient shepherd who stood just under five feet tall and sported a giant disarming nearly toothless smile. He lived in a sod, wood and tin cabin at the edge of the lake, where, given his deeply lined face, he was born a century or two ago. Erich enlisted him as our guide and goodwill ambassador.

West (blue) and Rafa (yellow) on the upper Río Gashán. Photo by Erich Schlegel

We scraped along the rocky streambed, using our paddles to push ourselves along. I was thrilled to finally begin the source-to-sea descent of the Amazon River for which I'd spent the last four years planning and saving. The going was difficult, leaving me short of breath in the rarefied air. Rafa, at least 80 pounds lighter than me, floated much more easily in the shallows, but he was a team player and traveled at my slow pace. Almost immediately, Jeff and John fell behind. After about half an hour of ineffectual knuckle-dragging they got out of their kayaks and began to walk along the grassy shoreline, dragging their boats.

Late in the afternoon we came to a particularly steep run about 100 yards long. For Rafa it was mere child's play, but the added gradient and closely spaced boulders created a pining hazard—the possibility of my kayak getting wedged between rocks, with me in it. We decided to walk around the hazard.

The clear sky had given way to clouds and then sleet. Now snow swirled in the rising wind. During the easy downhill drag I lost my footing on the wet snowy grass and caught my fall with my hand smack dab on a small cactus. Rafa dropped his kayak and ran in the nothing air to help.

West, John and Rafa call it a day near a herd of llamas. Photo by Erich Schlegel

Rafa stands a bit over five foot and wears a permanent warm toothy smile. He hails from Mexico City and speaks perfect English punctuated with exuberant white-water slang. He's polite, diplomatic and permanently stoked. As he plucked thorns from my hand, Rafa told me about the time he was invited out to an expensive restaurant during Carnival in Brazil. He didn't want to appear uncultured, so when the waiter delivered some exotic *hors d'oeuvre*, unidentifiable to Rafa, he picked up one of the small, thorny balls and popped it in his mouth. Immediately, the thorns lodged in his mouth and throat. His hosts rushed him through the revelry of Carnival to the emergency room, where a physician spent hours poking around his epiglottis with a pair of tweezers, removing hundreds of tiny spines. Apparently, Rafa told me with a laugh, you're supposed to cut open the spiny balls to get to the tasty innards. Perhaps my pin cushioned palm wasn't all that bad.

Rafa and West in an Amazon blizzard. Photo by Erich Schlegel

We were still laughing and plucking thorns when Jeff and John came trudging through the storm. John was worn down to silence, and Jeff looked like death warmed over. His rapid breathing wasn't enough to supply his body with oxygen. He was a drowning man. The whiteout blizzard cut into our tired bodies, and it was time to call it a day. We left the boats with a herd of llamas near a large rocky outcrop and began the slow trek back to camp.

Our distance for the full day's effort was a little less than a mile and a half, and we still had to get back to camp.

With John supporting Jeff we started back, leaning into the icy gale. Erich wrapped his sleeping bag around Jeff, who was too short of breath to tell us he wasn't cold, merely hypoxic. Every few feet we stopped to allow Jeff to lean on his knees. Rafa took Jeff's other arm while I carried the gear. In the midst of this retreat our team mentor, Piotr Chmielinski, appeared out of the storm, then ran back to inform the rest of the team of our situation. The support crew had already broken camp and was just about to drive away.

When we staggered into camp the team launched into action like ants rebuilding a crushed mound. Everyone worked with a single-minded, unspoken efficiency. Jeff was shoved into the warm van and stripped of his paddling gear, and Piotr got a magical blaze lit behind a boulder in the stiff, cold wind.

Rafa and John escorting Jeff through the blizzard back to camp. West in background. Photo by Erich Schlegel

We were mere hours into the expedition and it was already time for a new plan. Jeff's altitude sickness was debilitating, and John wasn't trained up in a whitewater kayak. His reputation for grinding out miles of flatwater didn't convey to these conditions. The Río Gashán was going to take far more time than I planned. Rafa and the other Tigers were only here for two weeks and we needed their help for the more difficult whitewater farther downstream. So, we fleshed out a plan to run the roughest sections with the Tigers, then paddle the easier sections on our own after the whitewater experts were gone. In the morning, Piotr would drive Jeff down the mountain to recover from his altitude sickness and John would join the land crew for the time being. The Tigers and I would make a fresh start from the place where we left the kayaks, aiming to knock off the 29 miles to Lago Junín. It all seemed to make sense in what was left of our oxygen-depleted brains.

As soon as the sun peaked over the mountains on Day Three we hiked with Erich and Mauro back to the kayaks. The team for the day included me, Rafa, Juanito de Ugarte and Tino Specht. At first, it was a wet and rocky roller coaster ride of steep, bumpy sloughs, then the terrain flattened out to an altiplano, or high plain. The Gashán was channelized, with frequent footbridges and barbed wire fences blocking the six to eight-foot wide stream. Now and then, the creek spread out and grew shallow. While the light and strong Tigers were able to scooch along with little effort, my weight kept the plastic bottom of my kayak pinned firmly against the streambed. With paddle and knuckles, I dragged and pulled myself along while the Tigers patiently waited. More than once, I cut my hands on the rusted barbed wire while we went over and under the sheep fences, leaving streaks of blood on my face and jacket. The thin air and my poor conditioning left me heaving for air.

The Río Gashán. Photo by Erich Schlegel

Throughout the effort I never lost sight of the beauty of the high lonesome landscape. Snowcapped Andean peaks stretched far across the horizon. Llamas, vicuna and sheep roamed about easily, their predators long since decimated at the hand of man. We were well above the tree line so no vegetation obstructed our view. Erich and Mauro hiked along, talking and laughing easily as I huffed and puffed. The clouds took on all forms, shifting from moment to moment. Mare's tails, cirrus, lens clouds, happy puffy clouds, anvil storm clouds and jet-stream-whisps were all part of the gallery. The temperature spiked and dropped suddenly with the passing or clearing of a cloud, leaving me sweating one moment, then shivering the next.

Several hours passed before the first snow clouds rolled down from the continental divide to our west. With fluctuations in the temperature, we weren't sure if the dark clouds meant a thunderstorm or snow, and I hoped for a dry snow rather than a soaking. By mid-afternoon it was clear we weren't going to make it to the tiny village of San Francisco before nightfall, but we had no choice but to push on. Now and then, Erich asked Mauro how much farther, and Mauro's answers became a source of entertainment. He was raised in a world without calendars or clocks, where time was a fluid non-entity. Things happened when they happened, but Mauro did his best for his guests.

Mauro, our mountain guide. Photo by Erich Schlegel

Erich: *"¿Cuantas horas?"* (How many hours?) Mauro: *"Media hora."* (Half an hour.)
Two hours later.
Erich: *"¿Cuantos kilometros?"* Mauro: *"Tres kilometros."*
Four kilometers later.

Erich: *"¿Cuantas horas?"* Mauro: *"Tres horas."*

Three hours later.

Erich: *"¿Cuantos kilometros?"* Mauro: *"Cinco kilometros."*

There was no rhyme or reason to Mauro's distance or time predictions. He walked this route his entire life. There had never been a reason to quantify the distance from his home to the village, or how long it took to walk there. It took as long as it took. Why would anyone care about distance or time since these were factors over which no one has control? Our contrived western constraints have no relevance at the top of the Andes.

Tino launching over yet another barbed wire fence. Photo by Erich Schlegel

I found great comfort in Mauro's snaggletooth smile. His patient weather-worn face matched his baggy pants, denim jacket and traditional yarn cap. He strolled as we struggled. The Quechua people had centuries of adaptation for this high altitude, having lived above 11,000 feet longer than any other culture on the planet, including Himalayan Sherpas. The number and efficiency of his oxygen-rich red blood cells made a mockery of our anemic sea-level blood.

The narrow stream often became too shallow to float. Each time I ran aground I tried to knuckle-walk to deeper water, pushing against the bottom with my paddle blade, or failing that lumber out of the kayak and drag it 50 or 100 yards to deeper water. Late in the day, levering hard with my paddle, I felt the warranty expire on the shoulder surgery I'd had six years before.

Tino and West on the Río Gashán. Photo by Erich Schlegel

There was no mistaking the tearing sensation of the tendon giving way and the sharp pain that I knew would be with me for months. More important than the pain, I noted the lack of movement. The muscle responsible for pulling a paddle was no longer attached as it should be. Months later, an MRI indicated that a few frayed fibers were all that held my shoulder together. As quickly as I recognized the injury, I decided my only option was to ignore it and keep the news to myself. The rest of the day promised more dragging than paddling, so I could baby the shoulder and hopefully recover enough to continue the expedition. We only had another 4,000 miles or so to go.

Soon dark clouds descended, depositing big white snowflakes through the bony, calm air in what would have been an ideal setting, given a big fireplace, thick blankets, good company and a bottle of scotch. I sent the Tigers ahead since they could go so much faster. Erich and Mauro would walk next to me and we'd all meet up at the bridge, which according to Mauro's highly detailed and accurate reckoning, was only a half hour ahead.

Of course.

Mauro closing a gate as West portages yet another fence crossing the Río Gashán. Photo by Erich Schlegel

I was able to see the Tigers struggling through the Gashán river bed along the flat plain for the next hour or so, until the snow became too thick against the dimming horizon. Erich patiently explained to Mauro, eager to lend a hand, that I wanted to haul the kayak alone. There was a certain pride and sense of accomplishment in the struggle. Though a slow burden to the Tigers, at the very least I was going to haul my own gear.

Expedition planning is a desk-bound activity, and the organizational crescendo of the last few months left me precious little time for upper-body conditioning and none at all for my lower body. Dragging the kayak through the rocks, shallow water and snow caused a nice deep burn in my thighs, coupled with extreme fatigue and breathing like an asthmatic. Breath-catching breaks became more frequent, to the point that an uninterrupted 30-yard pull was cause for celebration.

The snow turned to sleet, then drizzle, then back to snow. Light gray clouds slowly drifted close to the ground, spackling the dark gray horizon, only to smash themselves against mountain peaks. A wondrous silence accentuated the barren beauty. It all felt oddly home-like. There's a quiet comfort in the timeless solitude of these mountains, perhaps due to the lack of predators or the slow pace nature imposes on those who exist in this stark and desolate place. Perhaps it was the lack of oxygen in my brain, but I felt pretty good in spite of my condition.

Tino doing the limbo under barbed wire. Photo by Erich Schlegel

Once darkness came, the temperature would drop and things would turn serious. The sun gave up its efforts for the day, though a dull lightness hung under the dark clouds. We had no flashlights, though the Gashán clearly showed our route forward. I could have left the kayak if we had to make some speed and hike out across the altiplano. The lights from Cerro de Pasco twinkled their reassurance that warmth and nourishment waited a few short miles away.

Out of our silent trudge, Erich called my attention to a set of headlights roaring across the plain towards us. The lights jumped over berms, veered left then right, swerved around at Warp 8 and finally skidded to a stop a few yards from where I stood in freezing knee-high water. It was some impressive off-road driving from Jennifer Maxfield-Boucher, the future bride of the team's whitewater manager, David Kelly. Jennifer opened the driver's-side door and told us to get in. David was riding shotgun and they'd brought along one of the whitewater Tigers, Simón Yerovi, to haul my rear out of the descending gloom of a frozen nightfall. During the kidney rattling drive to rejoin the support team, David regaled us about Jennifer's experience running four-wheel drive through the Baja. We had a winner.

Driving the Andean highlands. Photo by Erich Schlegel

David and Jennifer tracked us using the satellite transponder attached to my kayak's deck. My sister, Barbara, along with the rest of my family, monitored my progress from her home in east Texas. Without much fanfare or question, Barbara immediately filled the void where an expedition manager should have been assigned. As is normal for my family members, they don't usually wait around to be asked to take on a task. On her laptop from thousands of miles away, Barbara tracked the transponder on my kayak and another in David's truck, using them to direct David and Jennifer to my location via satellite phone.

That Barbara filled this critical role without prompting was no surprise, nor would it have been for anyone in my family to do so. We were raised to make our own way and, just as importantly, to look after one another.

Barbara quickly established herself as our manager and liaison with the outside world.

David and Jennifer recovering the Tigers at an Andean sheep farm on the Gashan. Photo by Tino Specht

A quiet moment on the Río Gashán. Photo by Erich Schlegel

Given that National Geographic wanted regular updates for their website but instructed us *not* to use the BGAN satellite broadband unit, once we arrived in

Peru (due to their Everest expedition running up a $45,000 bill in a single month), Barbara and I established a daily sat-phone call. She took my verbal reports and turned them into engaging blog stories and social media posts, and she helped me deal with other business. She took control of the expedition finances and established great relationships with our sponsors and banker. She changed the contact information on the website to come straight to her phone and, with my thanks and praise, started running everything but the boots-on-the-ground aspects of the expedition. Most importantly, Barbara became my confidant. I could vent to her about anything and anyone and get nothing but sympathy and good counsel in return.

Marking the spot on my GPS, we left the kayak at the river and crammed into the small pickup. Nestled between Erich and Simón, I shook violently from hypothermia as Jennifer showed off driving skills she honed in the Baja desert and over the many years living in Colombia, the truck bouncing and grinding through the dark. Jennifer and David dropped us at the village community center in nearby San Pedro De Racco, and then drove back into the dark to find the Tigers.

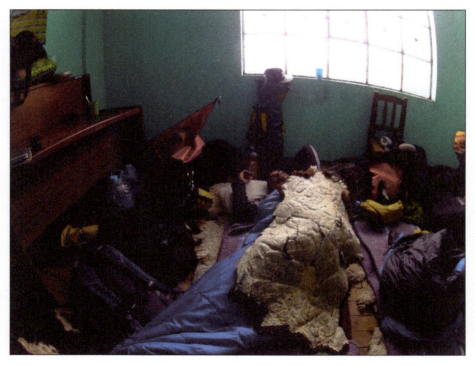

Warming up under lambskin. San Pedro de Racco community building. Photo by Tino Specht

The team zipped around while I sat in the heated van to thaw my frozen limbs and restore my core temperature, which had quickly dropped to hypothermic levels as soon as I stopped moving. John helped me out of my gear and into some warm clothes. My daughter Isabella, who developed the spreadsheets for our Amazon River environmental study, was particularly concerned about my well-being. In her 13 years on the planet she'd never seen me in such distress.

Soon David arrived with the Tigers, who'd found refuge in a farmer's one-room house. They too were hypothermic. The floor creaked under our weight as we hauled gear to upstairs rooms in the sparse but tidy community building, the use of which the village mayor generously offered. All but Rafa had to duck to keep from smacking their heads on the ceiling of the concrete stairwell. The building had no heat and scant electricity, but to us it was the Waldorf Astoria. After a warm meal, hot chocolate and just enough medicinal rum we rolled out sleeping bags on raw lambskins and clocked out.

Chapter Two
How, When and Why I Ended Up on the Amazon River

"If you're not ready to die today, then you better start living immediately."
EWH

My fascination with the Amazon grew out of my passion for ultra-marathon canoe racing. In 2008, adventure racing icon David Kelly and world-record distance paddler Carter Johnson were putting together a team to compete in something called the Great River Amazon Raft Race in Peru, but I didn't give it much thought. After all, it was in a foreign country and I'd never been outside of the United States. I knew Carter pretty well and raced with David in the Missouri River 340, where we set a new tandem speed record in 2010. I don't recall exactly how it happened, though I'm sure beer was involved, when David asked me to join the four-man team with him, Carter and legendary Hawaiian outrigger racer Mike Scales.

David, Carter, West and Mike. 2008 Iquitos, Peru

The Great River Amazon Raft Race is a 90-mile race that takes place over three days from Nauta to Iquitos along the Amazon River. Our pre-race prep consisted of eating as much as possible and drinking pint after pint of Iquiteña beer. After a bus ride from Iquitos to Nauta, then a prolonged celebration from the town leaders, a ferry dropped us at Fisherman's Island, where we and a crowd of racers from around the world were given a pile of balsa logs to build our rafts.

The local racers laughed as we spliced the eight 16-foot logs into four 30-foot logs to make a skinny raft similar to the slender canoes we race in North America. With the tropical blue paint still tacky in the humid air, we first sat on our skinny raft 10 minutes before the starting gun went off the next morning. The craft had a corkscrew twist and sat precariously low in the water. We looked like hippos next to our lean competitors. Even Carter, without enough fat to grease a skillet, seemed gargantuan by comparison. The top teams ignored us as freaks of nature.

When the gun went off we clicked our heels together and weren't in god-damned Kansas anymore, Toto. Though this was the first time the four of us formed a team, we did what came naturally after decades of racing, leaving a trail of fire on the water as we steadily crept passed the lead teams and smoked them all, winning by a record-breaking six minutes on the first leg of the race. Over the next two days we won the next two legs and finished with a new record which still stands at this writing, an hour ahead of the old mark. While the race was extremely exciting, what I got most from those two weeks in Iquitos was an appreciation for the Amazon River.

Never in my life have I ever seen something so large moving at such an incredible speed. I paddled the Missouri River and have seen the Mississippi on several occasions, but the Amazon dwarfed them both. Its size and power were unlike anything I ever experienced, and it captivated me. For the flight back to Texas, David lent me a tattered copy of *Running the Amazon*, Joe Kane's bestselling account of his 1985-6 Amazon expedition with Piotr Chmielinski. It took only four Jack and Cokes and a long layover in Phoenix for me to devour the epic story, which I never heard of until David handed me that paperback.

Simple as that, I thought. "Hell, I could do this."

I got off the plane and immediately informed my wife, Lizet, of my plan, which she regarded with the conservative solemnity with which she greeted all of my hare-brained schemes. My mother and sisters, never at risk of being mistaken for wallflowers, weren't quite as accepting. My mom fought me for months, but once she came on board, I never had a stronger ally.

Without the first clue of how to plan such a large-scale expedition, I went online to search for anyone affiliated with the *Running the Amazon* expedition. I quickly found expedition photographer Zbigniew Bzdak, who had gone on to a storied career at the *Chicago Tribune*. He made an email introduction to Piotr Chmielinski, who lives in Herndon, Virginia, an upscale D.C. suburb coincidentally named after the first North American to travel the length of the Amazon. Piotr graciously offered to video-chat the next day.

Just like that, I was in touch with one of the men who made one of the legendary descents of the Amazon river system, hiking, rafting and kayaking from its most-distant source all the way to the Atlantic Ocean. I was a little nervous to speak with this legend, especially since I didn't know how to pronounce either his first or last name. In spite of Piotr's patient tutelage and the best efforts of Beata, my uncle's Polish-born wife, my thick Texas tongue still has great difficulty with his name.

Piotr and I talked for two hours on that first video call, each putting away a bottle of wine. From that point on, Piotr was a gracious source for information and guidance throughout the four years leading to the expedition. Given the gift of gab, impeccable manners, regal good looks, academic intelligence and a quick wit, Piotr could sell air conditioners in the Arctic. In these abilities he is surpassed only by his lovely wife, Joanna.

Piotr had been masterful in navigating the tense team dynamics of an Amazon descent, though hard feelings still exist to this day. *Running the Amazon* describes an expedition coming apart at the seams, driven by internal politics, financial stress and the daily grind of extended wilderness travel. Their expedition had been organized by South African explorer Francois Odendaal, who dreamed of leading the first complete descent of the Amazon, including some 400 miles of whitewater on the Río Apurímac, which, at the time, was recognized as the Amazon's most-distant source. Odendaal assembled an international team of 10, including some of the leading whitewater paddlers of the day. Zbigniew and Piotr were invited on the strength of their experience with a group of Polish kayakers who talked their way out from under the Iron Curtain in 1979 to explore South America's most forbidding rivers, including Peru's infamous Colca Canyon. Kane, a writer who had never paddled a kayak, joined the team to chronicle the expedition.

His book takes us inside the expedition, as Odendaal, at turns overbearing and paranoid, loses the confidence of his team while Piotr quietly gathers influence. The team held together through the Apurímac whitewater, but couldn't long withstand the growing dissension amplified by constant danger and extreme discomfort. The group split when they reached the flat water below the confluence with the Urubamba River. Odendaal went home with most of the expedition funds, and the majority of the team went back to their lives. Only Joe and Piotr kayaked the remaining 3,800 miles to the ocean, with Zbigniew and team doctor Kate Durrant in support.

Piotr negotiating through another obstacle. Photo by Erich Schlegel

The Mother of all Monkey Wrenches

"If the highest aim of a captain were to preserve his ship, he would keep it in port forever."

Thomas Aquinas

The discord portrayed in *Running the Amazon* greatly influenced the way I assembled my team. I read everything I could find about the Amazon and expeditions in general, and all those books lead to two inescapable conclusions: there can be only one leader and that leader had to rely on the counsel of others. To that end, I made a point to invite only people I trusted. While I'd seek input and encourage discussion, the weight of every final decision rested on my shoulders. The survival of Shackleton's party after two years stranded in Antarctica validated that approach, in spades.

To set the arrangement in stone, I raised funds for the entire expedition, including all travel, boats, lodging, gear, food and essentials for all team members. Every penny would come from my own pocket, donations or sponsorships I arranged. I also made it clear to any potential members that they were there to support my effort to paddle solo from source to sea. If they didn't like that, or if they didn't fit in for any reason, I reserved the right to send them packing.

I applied for a $35,000 expedition grant from the National Geographic Expedition Council and was thrilled to receive an invitation to make my case at the organization's Washington D.C. headquarters. Gregg Treinish helped grease the skids with National Geographic via his organization, Adventurers and Scientists for Conservation. I'd already jumped at the chance to take scientific samples and make observations for Gregg's group during our trip. Beyond the potential grant money, I knew that any affiliation with National Geographic would open doors to other sponsors.

With a kayak strapped on top of my Subaru Outback I drove to D.C., stopping at rivers along the way to hone my rusty whitewater skills. Piotr and Joanna welcomed me into their home, and I had a very positive first meeting with the brass at National Geographic. Piotr sat in on the meetings, which I saw as a wonderful show of support. At the time, I had no idea how closely he was affiliated with National Geographic or how friendly he was with the organization's senior management.

My professional relationship with the State of Texas was taking precious time away from my efforts to raise funds, court sponsors and organize the trip. The entertainment I found poking holes in the bureaucracy was no longer all that challenging and my efforts in pointing out shortcomings in the various systems weren't appreciated by those in charge, so I unfriended my cubicle for greener pastures. To make ends meet I dusted off my old construction business and also worked for my family's social work firm, helping elderly people stay out of nursing homes and funeral homes. I also ramped up my fundraising efforts. No donation was too small, and I funneled every dime our household budget could spare into the expedition account.

That spring of 2012 was a blur of planning and activity, with the August launch date coming at me like a runaway freight train. During this time, I received an email from a whitewater kayaker named James "Rocky" Contos, who was planning an expedition to paddle each of the five principal tributaries of the Amazon River. Operating on a shoestring budget, like most explorers, he wanted to join my team on the Río Apurímac, which was the original starting point of our expedition. Rocky would piggyback on my logistics train in exchange for whitewater kayak support. I told him I was open to the arrangement, so long as he could fit within our timeline.

While most whitewater paddlers fit somewhere along the young hippie-jock thrill-seeker spectrum, Rocky was 40 years old with a Ph.D. in neuroscience, of all things. In high school, when most kids were chasing girls and trying to buy beer, Rocky developed an interest in growing exotic fruits. He received scholarships from the San Diego Garden Club, among others, and studied horticulture at the University of California at Davis, where he found he had a knack for the hard sciences. He earned his Ph.D. and landed a plumb postdoctoral position in the lab of a Nobel Laureate, where he soon discovered serious flaws in the research of the postdoc he replaced. His diligence eventually led to an investigation and caused his boss, the Nobel prize winner, to retract three papers she'd co-authored based on the flawed research. While the facts were on Rocky's side, his strident presentation torched bridges throughout the scientific realm. He hasn't worked in neuroscience since.

Rocky turned his energy to kayaking, with an insatiable appetite for exploration. He's notched more than 120 first descents in Mexico alone, and many more in Central America, Peru and around the world. While most whitewater boaters look to be the first to run the big Class V rapids and care little about the rest, Rocky liked to run rivers from top to bottom. He did the Class V stuff, plus all the minor rapids, briar patches, gravel bars and mangrove swamps. He often paddled alone, which might be due to the strange and difficult missions he chooses, or perhaps his personality. In any case, he approached expedition kayaking with scientific zeal and an eye for detail.

To prepare for his five-river, six-month headwaters extravaganza, Rocky spent months studying and annotating maps of the Amazon's principal tributaries. This process led to a monumental discovery: the Río Apurímac, which for decades had been considered the most-distant source of the Amazon and was at least 48 miles shorter than the neighboring Río Mantaro. Rocky checked his measurements five different ways and got the same result. The conclusion was rational and irrefutable: If measured from the most distant source to the sea, then the true source of the Amazon is not the Apurímac. It is the Mantaro.

If Rocky was right, everyone else had gotten it wrong. Explorer Loren McIntyre, who in 1971 traced the source to a tiny stream called Quebrada Carhuasanta high in the Apurímac drainage, got it wrong. National Geographic and the Peruvian government, who funded McIntyre's expedition and heralded his discovery on the cover of its magazine, got it wrong. The National Geographic Institute of Peru, which named a lake near the Apurímac headwaters in McIntyre's honor, got it wrong. A whole cottage industry of geographers who came later, vying to pinpoint the precise location of the source, but never looking outside the Apurímac watershed, all got it wrong. The list of honorable explorers content with a cursory examination of the question of the source even included Jacques Cousteau, who was drawn to the Apurimac source and noted in the book about his Amazon River expedition, that, *"Technically, the source of a river is generally recognized to be the farthest point of origin, not the largest of its headwater tributaries."* Through no fault of their own, Francois Odendaal, Piotr Chmielinski and Joe Kane got it wrong too.

If Rocky was right, the first descent of the world's largest and longest river from its most distant source was still on the table.

The Amazon from Source to Sea 31

Sowing the Seeds

"Defining the source of the Amazon is like unwinding a ball of string and trying to decide which of the tiny frayed threads at its core is, in fact, the end. By generally accepted definition, the source of a river is that tributary farthest from a river's mouth (as distinct, say, from the tributary carrying the greatest volume of water)."

Joe Kane, *Running the Amazon*

On April 2nd, 2012, just four months before the launch, Rocky shared with me his discovery, after I agreed with two stipulations: First, that I would keep the information secret, and second, that I would not make any changes to my expedition without his approval. It was a strange request, but I reluctantly agreed. Rocky then told me that he found the true source of the Amazon River.

We discussed his findings in detail, then followed up with a series of emails, where he asked for my written promise to the two stipulations. I thanked Rocky and assured him that his secret was safe with me. Any excitement I felt was eclipsed by my general lack of credence in Rocky's claim. After all, I was up to my armpits as chief bottle washer for a huge expedition, knew little about such geographic issues and was too engrossed in the legendary aspects of river exploration laid out in *Running the Amazon* to give much thought to Rocky's claim of another source. The known source of the Amazon had long ago been settled and agreed upon by luminary explorers and organizations such as the National Geographic Society, the Smithsonian Institution and Loren McIntyre. What evidence could a dirtbag kayaker (albeit one with a PhD) like Rocky bring to overturn such a pedigree? I politely put aside his claim and continued to organize the innumerable aspects of this rapidly developing downward-bound snowball.

Still, something about his description left a bit of an itch that I couldn't scratch. I logged onto Google Earth to compare the Mantaro and Apurímac, searching for holes in Rocky's theory, because if he was right and I stuck with my plan to start on the Apurímac, my descent of the Amazon would forever bear an asterisk. The more research I conducted, the more it seemed that Rocky had found something truly amazing. It seemed that people searching for the source over the last few centuries had given only a cursory glance at the Mantaro.

Still, the arena of source-naming was beyond my scope and I held little interest in moving my starting point based on a new and unproven theory. My original goal, which was soon to come to a seismic change, was to be the first solo paddler and the fastest to complete the Amazon.

I had no difficulty keeping Rocky's secret until April 19, 2012, when I received a text from Piotr, who said he had something important to discuss and would call in about an hour. Piotr rarely initiated our discussions, though he was always available for my pestering questions, so this call was a bit alarming. Rocky told me he recently pitched his "new source" story to *Canoe & Kayak* magazine and *Nature*. I knew that Piotr was close with *Canoe & Kayak* editor Jeff Moag, so I suspected he asked Piotr for comment given his legendary status on the Amazon.

Keeping my mouth shut about the source was one thing, but outright lying about it to Piotr—a friend and mentor who I respected deeply—was another thing altogether. I called Rocky and told him I suspected Piotr may know about the new source and that I did not want to lie to him if he asked me about it directly. Rocky said he would contact Piotr with the information, but I was to hold the company line. No problem.

Piotr's big news turned out to be something completely unrelated to the source, but before we hung up I told him to check his email for some information from Rocky Contos, who Piotr said he didn't know.

To my surprise, instead of blowing Rocky's claim out of the water, Piotr gave it quite a bit of credence, though he indicated, without any details or references, that the question of the Mantaro had been addressed in the 1980s. Still, after a long discussion with Rocky, he wanted to take a look at the supporting evidence. This was my tipping point. With someone like Piotr taking the claim seriously, I figured there might be fire under this smoke. Rocky, Piotr and I traded emails and spoke at length about the subject, and I researched river sources like a madman, reading everything I could find on the origin points of the longest river systems on each of the seven continents: the Nile, Missouri-Mississippi, Onyx, Volga, Yangtze, Murray-Darling and, of course, the Amazon.

I soon discovered that there is no *universally accepted definition* of what constitutes the source of a river, or the mouth of a river for that matter. Here we had centuries of scientific expeditions risking life and limb, traveling through uncharted hostile terrain to find the source of the largest and longest rivers on the planet, and no one had bothered to define what is meant by a river's "source." It's as if the French Academy of Science just tossed out an arbitrary stick and said, "That's a meter." In every discipline of science and engineering, standards are established in order to provide accurate and rational comparisons, without which the fine people of Texas could claim El Capitan as the highest peak on the planet, because it is two million cow patties tall, whereas Mount Everest stands at a mere 29,029 feet.

In short, Rocky was right. That meant I would have to move my expedition starting point some 500 miles and run a completely different river that, incidentally, no one had ever been down before. The primary goal of my expedition would also change, from a supported solo descent to the first *complete* descent of the world's largest river—and now, it seemed, also its longest.

As promised, I sought and received Rocky's permission to move the start of my expedition to the new source. In his written response, he said that it had been his desire to be the first to paddle the entire Amazon from its newly discovered and most distant source but given his time constraints he would "grant me the glory" of doing so. The wording was weird, but also generous. I was, and remain, truly grateful.

Rocky previously applied for a National Geographic grant for his five-rivers expedition, but the application had gone nowhere. Now, Piotr began using his influence as a personal friend of National Geographic leadership. News of a potential new source, together with Piotr's persuasive lobbying, quickly pried open the organization's checkbook, and a $7,500 grant came through in less than a week. The money supported Rocky's initial exploration of the source, but also seeded a good bit of discord.

The grant made Piotr the official leader of Rocky's "source" expedition, which didn't sit well with Rocky. More importantly, Rocky had to sign over to National Geographic all rights to publicize the discovery, from which Rocky drew the reasonable impression that *National Geographic* magazine would publish a story about the new source of the Amazon and him as its discoverer.

Piotr and his son Max flew down to meet Rocky and his two paddling teammates in Peru. They spent a few days searching the Andes above Lago Junín for the precise location of the source, eventually placing it at the apex of a high horseshoe ridge called Cordillera Rumi Cruz, from which emanated the Río Blanco and nearby Río Gashán. From there, a drop of water would travel farther to the sea than any raindrop in the entire Amazon basin, or the world.

After helping with the measurements, Piotr and Max returned home, leaving Rocky and his paddling partner James Duesenberry to make the first descent of the Mantaro, down some 500 miles of uncharted whitewater including dozens of

Class V rapids. Rocky would spend the next six months in Peru, making his way on a shoestring down the five major feeder rivers of the Amazon. He maintained meticulous GPS tracks of each descent, which would become the sixth and most definitive method of measurement he used to confirm his discovery of the new source. After Rocky had independently verified the new source of the Amazon, he came across an article in the May, 1991 edition of *South American Explorer* by Loren McIntyre. The magazine gave McIntyre free rein to say whatever he pleased, and he took the chance to settle a few old scores and muse about the fleeting nature of river sources. At the time, the Peruvian government was considering a plan to divert the Apurímac headwaters to the Pacific, after which, McIntyre wrote, "...it will be time to name a new 'true source' tributary. I have a candidate: the Río Mantaro." Remarkably, he identified the same source location Rocky would name 31 years later, noting that "...of freshets that renew Junín's waters, the most distant appears to be one that descends from a little known range of snow-peaks 20 miles from its northwest shore, the Cordillera Rumi Cruz."

Loren McIntyre on Mt. Mismi. Photo from the cover of the South American Explorer's Journal, 1991.

Just as he had with Rocky, Piotr quickly volunteered to be part of my expedition and helped coordinate things with National Geographic and offered to meet my team in Peru. With him, I met again with National Geographic to secure my own Explorer's Council grant and talk with various entities with the magazine and television station. The potential new source perked their interest, and my grant, too, was suddenly on the fast track. Like Rocky, I signed over all media rights to National Geographic and agreed to keep the new source secret, even while providing regular blog posts for our expedition website. I was happy to oblige. I had raised a good chunk of money, assembled a first-rate team and had the unqualified support of my family, legendary Amazon explorer Piotr Chmielinski and the National Geographic Society.

All was well.

Chapter Three

Back to the Very First Day of the Expedition

> "Oh, yeah. Oooh, ahhh, that's how it always starts. Then later there's running and screaming."
>
> Dr. Ian Malcolm

I was really concerned about Jeff. With all the duck-herding and minutiae that comes with wrangling a team to the top of the Andes Mountains to launch an expedition down the world's longest river, I hadn't realized how much the altitude would affect the team. I couldn't sleep. Actually, I *could* have slept if not for the incessant dry hacking emanating from Jeff in the next tent over. It must have been really unnerving for John, in the same tent, though obviously much more so for Jeff. The howling winds arrhythmically drumming our tents weren't enough to keep my wife, daughter and me from a wonderful snuggled up snooze, but Jeff's coughing was more worrisome than any passing storm. I'd raced marathon canoes with Jeff for 20 years and seen him in all kinds of hurt. We cemented our friendship as teammates in the Texas Water Safari, a 260-mile nonstop canoe race that takes the best and worst of a person and lays it bare for all to see. We both learned long ago that when something goes wrong there's nothing to do but keep going, so I had no doubt about Jeff's frame of mind. I knew he wouldn't quit. I just couldn't be sure he wouldn't drop dead. His cough showed no signs of abating, even as he munched fistfuls of coca leaves and took double doses of altitude pills. All night long he hacked that crusty dry autumn leaf cellophane hack.

At daybreak we crept from warm sleeping bags and unzipped frozen nylon walls. In the wild stillness, we spoke in whispers against the backdrop of the Andean highlands that stretched in all directions like a day-lit planetarium. The overwhelming silence was a far shot from the scene we'd encountered when we arrived just a few hours earlier. Our caravan of two trucks and a minivan rolled in well after dark, beneath a blanket of clouds and brutally strong cold winds. We circled the vehicles and piled bags of gear to block the gusts, and scrambled to set up our tents, bending stakes in the frozen rocky ground.

John (background) keeps an eye on Jeff, dealing with soroche. Photo by Erich Schlegel

The morning took a turn for the worse when my 13-year-old daughter, Isabella, threw up her breakfast. I immediately took this as a sign of altitude sickness, though in hindsight it had more to do with me forcing her to eat more breakfast than her stomach would allow. Bad father. Bad expedition leader. In my misplaced concern, I insisted Isabella stay behind as we paddled three miles across the lake to the newly discovered source of the Amazon River. That meant my wife, Lizet, would also have to stay behind.

After breakfast Piotr appeared in camp with Mauro in tow. He introduced everyone and explained who we were, filling the old man's gnarled hands with the miniature bottles of Jack Daniels I collected on the flight from Texas. A few minutes later Mauro's daughter, wife or granddaughter came over to introduce her baby. The mix of Quechua with Spanish was far too involved for me to follow, but the baby was all kinds of cute with round, rosy cheeks and the small dark eyes of her highland ancestors.

A frosty morning at the headwaters of the Amazon. Photo by Erich Schlegel

The morning was a flurry of activity, with David and Piotr reviewing the maps for our lake crossing and the short ascent to the source, and whitewater leader Juan Antonio "Juanito" de Ugarte organizing the raft and kayaks. Erich bounced hither and yon, snapping photos. Lizet and Jennifer took charge of camp logistics and the youngsters, Isabella and Alex, helped wherever they were needed. The lone grain of sand in the smoothly running expedition machinery came when John took a cup of hot chocolate from Jennifer without thanks, and then commented that it wasn't warm enough.

Simon with fossils he found at the source. Photo by Erich Schlegel

A shepherd on the banks of the world's highest lake. Photo by Tino Specht

Piotr and David plan the route to the source. Alex and John in the background. Photo by Erich Schlegel

We launched after lunch, four of us in the kayaks and eight more in the raft, including Jeff, who looked like death warmed over. Piotr led the team to the far shore of the lake, where the young whitewater Tigers quickly climbed toward the continental divide.

Piotr and West discuss what defines a river's source.
Photo by Erich Schlegel

The slope was littered with rough lichen-covered rocks, submerged in loose soil and tiny ichu grass, broken up by fissures of trickling water interrupting the enormous silence like low-pitched wind chimes. As the Tigers bounded ahead, Erich, Piotr and I made a more studied trek, stopping occasionally to admire the view and gulp lungfuls of thin air. Along the slender ledges off the surrounding steep mountain slopes, strings of llama herds easily made their way from pasture to pasture. Jeff and John remained on the shore, bodies prone and chests gurgling. After a scant hour of surveying, Piotr announced a three-foot hole in the dirt was the spring he and Rocky had identified as the true source of the Amazon, but then went into a long soliloquy that included numerous commas, semi-colons and asterisks, all amounting to something having to do with the flow going underground, then reappearing. We'd come at the start of the Andean dry season, and the trickle coming down from the divide appeared and disappeared several times on its way to Lago Acucocha, so Piotr proclaimed the source had to be the spot where the rivulet entered the lake at an elevation of 14,834 feet. Rocky would later identify a source about four kilometers away, at the headwaters of the Rio Blanco, at over 17,000 feet elevation, which he detailed two years later in a peer-reviewed paper published in *Area*, the journal of Britain's Royal Geographical Society.

Regardless of Piotr's winding ad hoc take on the subject, the lake was the first place you could float a kayak on what would become the world's longest and mightiest river. At 3 p.m. the crew gathered and I announced the official giddy-up for the expedition.

Rafa, Simón, West, David, Jennifer, Tino, Isabella, John, Lizet, Jeff, Juanito, Piotr and Alex. Photo by Erich Schlegel

Andean sunsets are as abrupt as the sunrises, so with long shadows cast by the surrounding apus (mountain gods) the team got in some last-minute mugs for the camera, held a perfunctory toast from a shared quart of local beer and started paddling back to camp. The kayaks let the raft get ahead, and Rafa handed Juanito a bottle—the latest exhibit in their ongoing debate about the relative virtues of Mexican tequila versus Peruvian pisco. Employing the scientific method, the two frequently held experiments to prove their respective points. Tino had a sip or two and I uncharacteristically kept my testing to a minimum to avoid antagonizing my omnipresent altitude-induced headache, which hinted at full blown *soroche*. Juanito wasted no time making up for my lack of enthusiasm, and most of the bottle was gone before we reached the middle of the lake. Rafa hung with Juanito in his customary role of sober young diplomat as Juanito's progress slowed and his navigation across the smooth lake proved rather erratic. Well after the rest of us landed, changed into warm clothes and supped on hot food, Juanito and his escort finally made shore. I lifted ragdoll Juanito from his kayak, then walked behind him as he staggered up the steep bank, laughing and committing wayward philosophy. More than once he embraced me to sincerely apologize for his altered state, though I assured him all was well.

The remaining Tigers, Simón and Tino, huddled in Jennifer's truck, firing up the heater now and again to ward off the biting cold. The sun dipped below the cordillera and darkness was creeping across the altiplano, bringing to life squadrons of unfamiliar constellations. Rafa poured the reluctant Juanito into the truck, but he wouldn't stay. Ten minutes later, he and Simón stumbled out, Juanito still wearing his paddling gear with his kayak skirt hanging from his skinny waist.

Juanito pursued Simón a short distance, taunting his taller and substantially more sober compadre. Juanito's gesticulations indicated alcohol-induced logic was being heaped upon some point of contention between the two. Fortunately, the ever-smiling Simón was playing defense, staying just out of reach. Keeping a close eye on the interaction from a distance, I pulled Rafa aside for an update. Seems shortly before the expedition, and well after Juanito enlisted Simón to pilot the raft, Juanito's girlfriend had left him for the laconic Simón.

The world of high-stakes international whitewater kayaking is a small community of extremely fit and ridiculously attractive young people. While their passports are full of exotic stamps, the desolate locations of their adventures and celebrations leave a pretty small pool from which to form romantic relationships. This leads to awkward situations when one relationship ends and another begins within the group. Yet the members of this elite fraternity trust their lives to one another every day on the planet's most challenging whitewater. There's no room for petty jealousy on such rivers. Juanito knew that as well as anyone. The next day, he would sober up and focus on the job at hand: leading a team of experts and novices through 500 miles of powerful whitewater, with a raft pilot who was doing the belly-polish with his ex-girlfriend.

Base camp in the Andes. Photo by Erich Schlegel

Our fire burned low behind the boulders sheltering it from the rising wind. Hypoxia, while guaranteeing fatigue, also prevented sleep, so one or two of us stayed up to talk about nothing. Jeff wheezed and coughed worse than ever in the tent he shared with John. My girls bedded down early while Erich and I contemplated the stars.

By now, I had only scratched the surface of the centuries-old debate regarding the source of the Amazon, so I relied on Piotr's leadership. At the heart of the academic issue was the subjective definition of what constitutes a river's source. In the months since Rocky shared his news I'd learned that the established sources of the world's most famous rivers fit a variety of definitions, but all hover around the idea of "most distant." Still, it was all conjecture with the numerous variables added to benefit whomever was planting their flag for a first descent or ascent. I wasn't the first to stumble upon this debacle, which plagued Burton, Speke and Livingstone at the Nile, and little did I know, as we launched the 4,200-mile expedition that I was already up to my waist in the political sludge that included National Geographic and centuries of Amazon River expeditions.

For now, I had an expedition to run.

Isabella and Alex with Mauro's family. Photo by Erich Schlegel

Chapter Four
The Wild, the Innocent and the Peruvian Shuffle
Day 4, Mile 16.8

> *"The muleteers I had engaged were drunk at an early hour and I had to send the police after them. It is really curious to observe how entirely indifferent to the fulfillment of a promise these people are, and how very general the vice is. These muleteers had given me the strongest assurances that they would be at my door by daylight, and yet when they made the promise they had not the slightest idea of keeping it. The habit seems to be acquiesced in and borne with patience by even the true and promise-keeping English. My friend Mr. Jump did not sympathize in the least with my fretfulness and seemed surprised that I expected to get off."*
> William Lewis Herndon, Cerro de Pasco, 1851

All was well in San Pedro de Racco under the shockingly blue crystal cold clear sky of morning. The agenda for the day was to rejoin Jeff, Alex and Piotr in Tarma to figure out how to get through the Mantaro's 500 miles of uncharted rapids in the next two weeks. Sixteen miles in four days wasn't going to cut it. This was far slower than my most conservative estimates and underlined one of the most important things I'd learned about expeditions: Predetermined timelines are irrelevant.

Our minivan and truck caravan stopped in the town of Junín for the two-hour Tiger feeding at a nice café overlooking Lago Junín. In Quechua, the lake is "Chinchaygucha", which translates rather unpoetically into "northern lake." The Spanish name "Junín" came after the famous Battle of Junín, fought on the marshy plains surrounding the lake in August, 1824. Though little more than a skirmish, it was a great moral victory for Simón Bolivar's patriot army, who routed the Spanish cavalry and sent them running for the safety of their infantry.

A flamboyance of flamingos on Lago Junín. Photo by Tino Specht

Lago Titicaca, along the Bolivian border, gets most of the attention, due to its large size and navigability, though it sits about 1,000 feet lower in elevation than Lago Junín. The lake itself is unremarkable, except for its status as the highest large lake in the world, at an altitude of 13,395 feet and covering more than 200 square miles. Migratory flamingos inhabit the swampy grasslands that stretch for miles around the lake, which is protected as a wildlife sanctuary. If ever you want the rare opportunity to use the grammatically correct "a flamboyance of flamingos," this is it. We witnessed several large flamboyances among the reed-clogged banks of the lake, and I can attest the flamboyances were indeed flamboyant. Waves of less flamboyant birds flushed across the skyline, taking off and landing in huge groups that, no matter their size or beauty, will never be anything but "flocks," though I must admit to glimpsing a murder of crows later on, or was it a conspiracy of ravens, I can't be sure.

A couple of hours after lunch, we stormed back into the beautiful valley, where Tarma lay nestled amongst neatly terraced layers of flowers, cultivated to fine, vibrant colors to be shipped around the world just in time to grace a dinner table or capture a loved one's heart. Amidst a flourish of dust and rejuvenated by what felt like thick air at 8,000 feet elevation, the entire team took the Hacienda Santa Maria by storm.

The pressing issue was our miserable progress thus far, and the limited timeline for the Tigers. Piotr suggested we divide the Mantaro into sections, and run the most difficult stretches before Rafa, Tino and Simón left the expedition in two weeks. I could then double-back to run the easier sections with Jeff and John, or alone if need be. The plan wasn't ideal because it meant we would not make a continuous top-to-bottom descent of the Mantaro. However, it still accomplished the goal of paddling the entire Amazon River from its most distant source. More importantly, it was the only good option left.

Scouting the Rio Huarpa confluence. Simón with John and Piotr in the background. Photo by Erich Schlegel

Juanito agreed with the approach. That night he, David, Piotr and I unrolled the maps and divided the river into seven principal sections. The first section included the canyons just south of the highland metropolis of Huancayo, down to the Tablachaca Dam. Most of Rocky's team ran these sections, so with the help of his journal notes and my topo maps we had a general idea where the danger lay. About 60 miles downriver of Tablachaca, the Río Huarpa joins the Mantaro, which then disappeared into the middle canyons. These canyons would be our next section with the Tigers, down to the Chyquia Bridge. The lower canyons, from Chyquia to the cloud forest, would be the final difficult stretch. After safely leading me, Jeff, John and Erich through the rapids, the Tigers would depart and we'd paddle the easier sections from Lago Junín down to Huancayo, then Presa Tablachaca to the Río Huarpa confluence on our own.

Tarma valley. Photo by Tino Specht

Most large-scale Amazon expeditions have a "fixer" in Peru, whose job is to handle and smooth over all the logistical nightmares involved with shipping kayaks and gear into the country. The customs offices are quite complicated and most of the requirements for importing needed supplies will not make sense to anyone from anywhere else, nor to most Peruvians, for that matter. Our fixer was Teresa, whom I met through Piotr. Teresa grew up in Lima, but now lived in Wyoming, and previously worked with Peruvian Customs. When another Lima contact, Alvaro, refused to pick up the solo sea kayak from customs as he had agreed to do months before, we had 30 days to get the kayak out of storage before it would be sold at auction.

Teresa quickly confronted Alvaro about his excuse of not having a business license required to accept the shipment, to which he admitted to simply not wanting the liability for the kayak. We had shipped the kayak in Alvaro's name at his request because he said that was the only way he could get it out of the customs warehouse.

Working the phones from Wyoming, Teresa arranged for her uncle Willy to pick up the kayak. Teresa miraculously managed to switch the recipient's name to her uncle's name from several thousand miles away. I was impressed. She knew the system and still had contacts in the bureau and at the last minute the kayak was surreptitiously transported from the customs warehouse to the patio of Willy's house.

As Teresa worked to spring the kayak from customs, I asked several times, in writing, if she was planning to charge me for her efforts, and if so, how much. She responded within minutes to all other emails, but none of my inquiries about money. Any lack of response in this day of immediate electronic communication usually leaves me a bit suspicious, but I figured that my offer of money for a favor had somehow insulted her, and she was graciously ignoring it.

I was sold. Teresa was a great addition to the team. Her energy and depth of knowledge were a welcome addition to what I felt had been a frustrating one-man operation getting all these parts in motion. She and I talked about getting the two tandem sea kayaks from Nottingham, England to Peru. Once again, I needed a Peruvian citizen to accept the kayaks in Lima. There were several other restrictions, such as getting a special bond ($5,000) from a Peruvian bank, which the customs department would hold until the kayaks were officially deported from the country. This was to prevent me from selling the kayaks in Peru, which would create unfair competition with Peruvian kayak manufacturers or dealers, of which there were none.

I agreed to pay for Teresa's flight to Lima, where she could handle all this in person. Having just purchased all the team airline tickets, I knew she could fly round trip from Wyoming to Lima for around $900, which seemed like a good investment. Teresa would get a vacation with her family in Lima and we'd get the kayaks. Everybody wins.

Teresa told me to ship the two tandem kayaks in her name to Willy's house. Jason, the co-owner of Valley Sea Kayaks in England, drove the kayaks across Nottingham himself and watched them being loaded onto the ship to Lima, Peru.

Piotr suggested I offer Teresa $500 for her help with the expedition, which I did in an email. She answered that she had already spent $800 on phone calls to Peru, which *"would barely cover the expenses I already accrued in helping you out."* This was surprising, since I was calling Peru daily and never had a phone bill above $200. To her thinking, she was doing me a favor and I should expect to pay for it, even without any prior agreement. It would be a few more weeks before I realized the line between a favor and a business transaction had been crossed without my knowledge, and history proved that I was in good company.

In his book, *Walking the Amazon*, Ed Stafford writes of an old woman who invited him and his two hiking companions into her home for tea and a hot meal, then offered to let them pitch their tents outside her door. The evening was thick with smiles and hospitality, all of which was shattered in the morning when the woman told them they owed an enormous sum of money for the food, tea and camp spot. From then on, Ed always asked, "How much?" before accepting any favors in Peru.

While there is no malice in what became known amongst our team as the "Peruvian Shuffle," there is most certainly a deficiency of altruism. I was raised in a culture where an offer of money for a favor is always declined, and the terms of a deal are agreed in advance. That's just not how business is done in Peru, and it's been that way for a long time.

In 1850, U.S. Navy Lt. William Lewis Herndon was dispatched to survey the Valley of the Amazon and make a report to Congress, not unlike Lewis and Clark's exploration of the North American interior a generation before. Born in Virginia in 1813, Herndon went to sea at 15 and earned his commission at 27, later commanding the brig *Isis* during the Mexican-American War. Starting with a party of six men in Lima in May 1851, Herndon and Lt. Lardner Gibbon spent 11 months crossing the Andes and surveying the Amazon basin with eyes steeped in the doctrine of manifest destiny. Herndon's wonderfully detailed and illustrated report became a popular sensation. His prose brought the mountains and jungle to life, sparking the imagination of fortune-seekers and would-be adventurers around the world. Among them was a young Samuel Clemens, who promptly traveled to New Orleans to take passage to Brazil, but he couldn't find a ship and signed on as a deckhand on a Mississippi riverboat instead. If not for that accident of fate Jim and Huck may have ventured down the Amazon rather than the Mississippi.

Herndon put his pen to the task of describing the Peruvian Shuffle in all its variations. The 37-year-old Lieutenant knew the dance well by the time he arrived in the highland village of Cerro de Pasco in 1851, but still wrote scathingly of the mule drivers who spent their advance on liquor and blew off the appointed dawn meeting.

Little had changed by 1970 when American adventurer John Ridgeway and his team hired some guides with a few mules and a horse to haul their gear on a 10-mile trek along the Río Apurímac. Midway through the hike the guides stopped and informed Ridgeway that the agreed price for their services had doubled—take it or leave it. Ridgeway pitched a huge fit, and when he finally calmed down he proposed a different deal. He would simply buy the horse and be done with the guides. They wanted thoroughbred money for the boney nag, but eventually a price was reached and the guides departed. Only later did Ridgeway learn that the horse had not belonged to the guides who sold it to him.

I certainly wasn't going to haggle with Teresa over the $300 difference between my offer for her services and the amount she now expected for her phone calls to Peru. Though a bit tweaked by her up-selling of my $500 gesture of thanks, I sent her a check with a kind thank you note. This was a small price to pay for what Teresa had accomplished in getting the kayak out of customs.

A few days later, I emailed to ask her preferred departure and return dates. She responded that she had already gotten her ticket and that I could just mail her a check for $2,000 to cover her purchase. Holy crap. Two thousand dollars! A quick online search revealed a first-class ticket was $1,600.00. I held my tongue and mailed another check, as Teresa steadily worked her way outside my circle of trust.

A few days before the team left for Peru I sent one of my regular email updates to all the expedition members. Amongst the routine reminders and logistical items, I added a tribute to Teresa's abilities, explaining what her duties would be in Lima and praising her abilities to work through the customs bureaucracy. At the end of the email, I joked that the customs officials had better do as she says or she would "haunt them" until she gets what she wants.

Teresa, included in the team email, responded with a note that I was on the hook for all her taxis, meals, tips, bribes, gifts and other expenses while she was in Peru. *"I hope I won't have to haunt you,"* she wrote.

Bam! The icing on the shit cake I was about to eat. I consulted Piotr, who recommended, once again, that I offer Teresa $500 for her time and expenses. I did just that, citing the tight budget we were under and our original agreement that she would get the kayaks out of customs in exchange for me only paying for her flight down to Lima.

Teresa didn't respond, but I learned from Piotr that she was gravely insulted and might just forget the whole thing, which would leave our kayaks to be confiscated because the addressee—Teresa—wouldn't be there to receive them.

And so there it was: the kayaks were in Teresa's name, on a ship in the middle of the Atlantic Ocean. I'd overpaid for Teresa's phone bill and airline ticket and now, with no possible way to get the kayaks without her, the dance was complete.

The only thing left for me to do was to apologize for the "insult" and mail another check. Then with gushing graciousness, appreciation and praise for her efforts thus far, I asked for a breakdown of what I should expect to pay for her expenses and services in Lima. Her itemized list came to an additional $1,400 and concluded with this gem: "*I am only charging less than minimum wage due to your low budget… Thank you!*"

Of course.

We met Teresa on the plane to Lima, where I gave her the cash and a gracious hug, and nearly bit my tongue in half. So, in the quaint mountain town of Tarma, in the beautiful courtyard of the historic Hacienda Santa Maria surrounded by bright spring flowers, when Jeff told me that I was expected to pay more money to Teresa, it didn't sit well. I was already into Teresa's favors for almost $5,000.

Teresa's uncle Willy, who invited expedition members to stay at his house upon landing in Lima, also expected payments greater than the five-star Lima Hilton and did not make this known until it was time for us to depart. Jennifer, fluent in Spanish and far from a shrinking violet, talked Willy down from the $150 to $100 for her group's four-hour stay.

Teresa and Willy's offers of assistance threatened to bankrupt the expedition at every corner.

During our lunch stop in Junín, I learned that our boxes of Spiz protein and carbohydrate powder had been discovered at a post office gathering storage fees. Willy's address and phone number were clearly printed on each box but the postal workers decided to use telepathy instead of contacting him. After several weeks in storage, the powder was about to be confiscated, so Willy allegedly paid $240 to retrieve the boxes, that would have cost far less off the shelf, and now expected me to pay him back. I decided to let Willy keep all the product, since he hadn't checked with me before paying the money. Given the option, I would've let the post office keep the protein powder.

Then Jeff, who was looking slightly less gray in the relatively dense valley air, told me that he paid Piotr $270 to reimburse him for the postal fees. Willy did me a favor and paid the fees. Teresa simply asked Piotr for the money. Piotr, doing me a favor, gave it to her. Jeff, doing me another favor, reimbursed Piotr. The storage fees were $240 and Teresa added another $30 for the $10 taxi ride to pick up the boxes. Apparently, the taxi ride wasn't included in the $1,500 I paid Teresa in advance to handle our logistics in Lima.

The expedition could no longer afford people doing me any favors.

Chapter Five
The Mantaro Pushes Back

Day 7, Mile 16.11

"The greater danger for most of us lies not in setting our aim too high and falling short; but in setting our aim too low, and achieving our mark."
                                                                    Michelangelo

From Tarma, we drove to Huancayo to re-structure for our launch into the white water of the Mantaro. The La Cabana café in Huancayo is a well-known gathering place for explorers, justifiably famous for its sausage pizza and signature hot tea made with local herbs and served in a tall carafe with an appropriate dose of rum, designed to ward off the cold Andean nights. The proprietor, Lucho Hurtado, is a renowned explorer, from whom adventurers from around the world seek sage advice and quiet connections, and I was no different.

Lucho's birthday party. Huancayo. Photo by Erich Schlegel

I met Lucho through Piotr (I'm convinced that Piotr knows every person on Earth) and he the entire team. The expedition was off to a tough start. I'd blown out my shoulder, Jeff was knocking on death's door, we were hemorrhaging money and we had to completely reformulate our game plan for the critical whitewater section—all in the first three days. There was plenty to worry about, but now the whole expedition joined Lucho's birthday celebration in his private saloon. Juanito taught my daughter, Isabella, Peruvian-style dance, while we all swilled champagne, ate like kings and shared a huge cake with Lucho's family and friends. It was an amazing night to remember. All expeditions should spend the occasional night in revelry. Temporary liberation from the weight of the world is simply fantastic.

The Mantaro River cuts violently through several mountain ranges along the Andes on its route to the jungle. One of these deep incisions begins just downstream from the town of Vieques, a suburb of Huancayo. This is the last possible launch point before the earth squeezes the free-flowing Mantaro into a chasm of swift-flowing energy forcing itself towards the ocean. Rocky called this landform Izcuchaca Canyon, after the town 40-odd miles downriver.

The Mantaro flowed swift, brown and smelly below the bluff on which Vieques was settled. We crossed the river on a substantial steel suspension bridge then drove to the southern outskirts of town, passing quaint houses with well-tended landscaping and effervescent squares, alive with music and celebration at the drop of a hat. We launched across the river from a brightly painted and polished water slide park, set surreally against a backdrop of huts heated by animal dung fires.

We stuffed the four kayaks and the six-man raft with far too much gear. Simón teetered on the back of the glorified yellow inner tube while John, Jeff and Erich squeezed between dry bags and camera gear. My kayak turned from a nimble sports car into a sluggish concrete coffin. The Tigers took pity on me and crammed even more gear into their kayaks, yet I still sat low in the water. I honestly did not know how I was going to maneuver through the big water just downstream.

We weren't sure how long it would take us to get to the next bridge, Puente Cucharan, about 20 miles downstream, so we packed for several days. This was the section where Rocky, months before us, guessed a wrong route at the top of a Class V rapid and lost his boat, paddle and most of his gear. The next day he climbed out of the canyon while his partner, James Duesenberry, continued downstream alone. Though Rocky found his kayak days later with the help of local authorities, the Mantaro had shown its teeth.

Juanito, concerned with the truly awesome quantity of camera gear Erich stacked on the raft, told him he'd have to leave some of it behind. It was nice to hand off a good chunk of leadership to Juanito, who directed the operations, checked and rechecked everyone's gear, and assigned duties and chores. I liked his leadership style. As author James Webb put it, he wasn't in the shit business—he didn't give it or take it.

Juanito hailed from Peruvian whitewater royalty, as it were. His father, Tony de Ugarte, is a legend who pioneered kayaking in Peru and was a National Champion in the 1980s. At his father's knee, Juanito learned to kayak and read water as an almost indescribable art. His understanding of moving water was so deeply ingrained that it took great effort for him to describe its nuances in terms we normal humans could understand.

Juanito was very diplomatic and matter-of-fact when confronting Erich about his gear. It was simply a matter of weight and space. The two of them went back and for in quick Spanish, but it was Juanito who held firm. Erich would have to prioritize, which he did, begrudgingly. He was a professional and wanted to have the best gear to photograph the expedition, but recognized that safety was paramount.

As Juanito and Erich haggled, I sat sweating in my dry suit, a Gore-Tex getup with watertight rubber gaskets at the neck and wrist. It had been the perfect antidote to the daily snow, sleet and rain near the source and on the Río Gashán. Now in the bright morning sun it was a tight-necked straightjacket, doing a fine job of retaining buckets of sweat. Though the water was still ice-cold, the air was warm and several team members stripped to t-shirts.

Squeezed into my Porsche of a kayak, I envied the guys in the raft, wearing minimal gear with all that room to move around. My knees were wedged against the padded sides of the tight cockpit and my feet, pressed hard against the foot braces, were already going numb.

As we found in the thin air below Lago Acucocha, a lot can happen in three kilometers. The Class II rapids were no big deal. Then we hit a few Class II+ and III rapids, which I approached with a bit more caution. Following the Tigers made it easier, but in one of the rapids I leaned my weighty kayak too far while punching through a small hydraulic and quickly tumped over. I opened my eyes in the clear cold water to see the river bottom shoot by as I set up to roll on the right, my strong side.

A reliable Eskimo roll is an essential whitewater kayaking skill, usually performed under duress. While hanging upside down with the drowning clock ticking away, the paddler reaches one paddle blade toward the surface and snaps the boat upright using the hips and knees, with the head coming out last. My technique had never been great, but I usually managed an ugly roll. Right now, that's all I needed.

Everything went well for about a millisecond. Then, halfway up, my injured left shoulder gave way. I felt the tendon I'd damaged a few days earlier rip even more.

Crap.

I flopped upside down then repeated my setup for the roll, holding my right arm lower in the water in hopes that I could throw a huge hip snap to compensate for my failing shoulder. I snapped my hips so hard they would've made Elvis blush, but barely rocked the boat. With my heart pounding and my last lungful of air used up, I pulled the handle of my cockpit skirt and bailed out, an inexcusable humiliation for a whitewater kayaker.

I came up cussing. My shoulder was on fire as I held tight to my paddle and clung to the grab-loop on the stern of my boat, guiding it through the rapid. With my feet in front of me, I bounced off rocks and dodged the worst of the whitewater until Tino arrived to haul my sorry ass to shore. He stayed with me in the eddy, offering support and consolation as I bailed the water from my kayak. I knew he was surprised at my inability to hit my roll in an easy rapid, but he was too cool to come out and say so.

Juanito and Tino spoke out of earshot as I wriggled back into the tightly loaded kayak, determined not to repeat this humiliation. For the next few rapids Tino was barely out of arm's reach. The young gun was in an uneasy position, having to watch the boss without letting on that he and the others had decided I was completely out of my element. I could have told them my shoulder was injured, but the shoulder was irrelevant. The only thing Juanito and the Tigers needed to know was that I couldn't roll. Tino made small talk as we continued through the Class II and III rapids toward the much more substantial whitewater downstream.

We caught up to the raft and learned that Erich and Jeff had been thrown from the raft when it plowed into a rock. It's not unheard of to buck a rafter in easy rapids. If that were the only issue we would have carried on, but Juanito was concerned. In the sunless depths of the canyon the mercury was dropping quickly. Jeff and Erich were dressed in summer clothes and showing early signs of hypothermia, so Juanito made the call for us to stop and get a fire going.

After spotting a camp spot, Juanito directed me to catch an eddy while the team unpacked the raft. It was time for a chat. My two roll attempts had apparently been so feeble that he hadn't noticed them. He thought I'd bailed out of the kayak without even trying to roll, an inexcusable breach of the whitewater code calling for a come-to-Jesus talk. Juanito was concerned that I couldn't roll and wanted me to show him that I could. My shoulder had only recently gone from a stabbing pain to a dull ache and another roll or two at this point wasn't going to help. Still, he was well within his rights to ask.

"Let's see it, man," he commanded quietly.

Damn.

I took a breath and rolled upside down in the calm clear water, easing my dilapidated shoulder into position. It felt as if only a few threads of connective tissue were still holding the joint together, and for a moment I wondered if I could roll without using my left arm. Not a chance, I thought. Just get it over with. With one clumsy, mule-like thrust I cowboyed a butt-ugly roll. My shoulder lurched afire, but I was upright again. I wiped the water from my face, using my hand to hide my grimace. Okay, let's pour the rum.

"Let's see that, again, but be more fluid." Juanito was in coaching mode.

Shit.

This time, I was going to hip-snap the crap out of this mother, to reduce the force needed by the paddle. Mind you, 50-year-old white guys don't have much hip to snap. I took another breath and tumped back upside down into the cold water, swinging the blade in a half stroke just powerful enough to allow my haggard hip snap to gain some purchase. It wasn't pretty, but I was upright again. I swallowed a yelp as the pain swept from my shoulder all the way down my left arm. Two for two; surely, we must be done now.

"Okay man, this time do it without thinking about it. Make it one fluid motion."

Goddammit!

This time I could barely bring my left arm across my body to ready for the move. I grunted and pushed. All cylinders were firing. My right arm swept the blade back and my hips snapped as they should, but the left arm remained motionless. I couldn't move it. My head came up first; a rookie mistake. I gulped air and then went back under. By then, the left shoulder had cashed in. I tried sweeping the right blade without moving my left arm, snapped with all my might and went absolutely nowhere. I grabbed Juanito's bow and wrestled myself back upright.

"Okay, man, you're tired. Let's call it a day." Juanito was merciful and respectful, but he found out what he needed to know.

Early camp in Izcuchaca Canyon. Photo by Erich Schlegel

We had a fire going quickly, and as we settled in its warm light Tino railed about the trash sticking to every rock and branch on the Mantaro. In fact, the huge Huancayo city dump, upriver, is located along the oft-flooded banks of the Mantaro. He launched into a glorious tirade against the assholes degrading mother earth and the river. When someone accidentally tossed a trash-covered branch into the fire, causing an emission of toxic black smoke, Tino stormed off into the dark, cussing. A few of us huddled next to the fire under a Buick-sized boulder, passing the rum while others bedded down early. The moment was quite wonderful after the air cleared, in spite of our slow progress. I was happy to be away from the city and on the move, albeit slowly.

The next morning, we gathered our gear and donned our river armor with an optimistic feel in the air. The raft team was a bit shaky and my roll was non-existent, but we were on our way. This was our first morning without a support crew and we were going to make some headway. Spirits were up as we gulped instant oatmeal and steamy mugs of coffee. Rafa would watch after me for the day. As the others loaded the raft, he and I paddled into an eddy and practiced some maneuvers, Rafa assessing just how rudimentary my skills were. The shoulder was stiff, but workable. The midnight hit of hydrocodone still circulated in my system so the full-bore pain had yet to resurface. This gave me some confidence that I would be able to execute a roll, though I was determined not to flip in the first place. Rafa was the perfect gentleman, giving me kudos for little successes without a hint of condescension, and his optimism was infectious.

With Juanito running the show, Tino on safety, Rafa watching my six and the raft secured, we pushed into the rushing Mantaro. The raft lumbered and maneuvered like an elephant ballerina. Tino zipped ahead to take photos and pick up anyone taking a swim. Juanito hovered here and there barking out directions, while Rafa and I bobbed in the shadow of a sheer granite cliff above the first rapid, waiting for the raft. Before it reached us, a slew of bowling-ball-sized boulders rained down from above.

"Let's go!" I called out as I blasted away.

Once below the rapid, looking back upriver, I saw two boys laughing and running from the cliff. *Sonsofbitches.* Rafa caught up just as the boys disappeared.

"Those boys dropped those rocks on us!" I yelled over the rapids. In his usual glass-half-full perspective, Rafa replied, "Nah, it was just a rock slide." His bright attitude nourished us all and helped lighten my darker side.

We ran a few small drops, then pulled over to watch the raft maneuver around a house-sized boulder. The rock split the river, offering a tight squeeze on the right and a wider 90-degree corner on the left. Most of the flow went left and there was plenty of room to make the move. The entire Mantaro slammed into this boulder, so hitting it was a really bad idea. Individually, each of the rafters had sufficient skill to drive the raft easily around the boulder. Of course, they weren't paddling individually; they were four men in one overloaded raft. They needed to work together, and that clearly wasn't happening.

Their paddles flailed in all directions, leaving the raft at the mercy of the river. Simón, stationed in the driver's position in back, stretched far out over the water, laying down powerful draw strokes in a futile effort to keep the raft facing downstream. Normally, a raft full of paddlers unfamiliar with the river and one another would eddy out before a rapid to talk about the route they planned to take and what was required of each paddler. It was strange that our rafters didn't do that, but I figured with someone as experienced as Simón at the helm, they surely must know what they were doing. The faster they flew into the rapid, the less sure I became.

Simón (red helmet), John, Jeff and Erich drifting sideways into a rock. Photo by Tino Specht

At some point in every debacle there comes a recognizable point of no return. Those of us watching this tragedy unfold instinctively noted the instant the floundering raft crossed that invisible threshold. With four paddlers churning hard in opposite directions, the raft zoomed towards the boulder. At the last second a couple of the guys muscled their paddles against the rushing current towards the thin right-hand channel, but it was too late. With the crew still pumping away, some clawing left and the rest working right, the raft slammed squarely into the massive boulder. For a fraction of a second it appeared that the raft would just stop, but then the inevitable happened. Like a giant yellow and blue slug under the force of the river, the raft began climbing the two-story boulder. The guys in front grabbed for ropes and tie down straps as their world went 90 degrees from normal and they gained altitude. In another half-second eternity they lost their grip and dropped onto their teammates in the back of the raft.

The stern quickly submerged, caught between the immovable boulder and the boundless force of the river. The raft flipped like a pancake. Simón rolled off his perch on the back, landing feet-first in the water. This catlike maneuver ended with Simón's foot lodged between two rocks on the riverbed, trapping him under tons of fast-flowing water. The Mantaro shook him like a ragdoll against the raft as he struggled to free himself. Gulping equal parts water and air, Simón wrenched his foot free and kept a death grip on the overturned raft.

Boulder: 1, Raft: 0. Photo by Tino Specht

The other rafters, having fallen free of the raft, swam like crazy around the boulder, kicking clear of smaller boulders and bumping through the rapid. The raft fell sideways off the boulder and shot downstream with Simón still hanging on. Only Erich was able to reach the safety line thrown from shore. Rafa hauled him to the left bank as Tino and Juanito raced downriver in their kayaks to rescue the remaining three. After we helped Erich ashore, Rafa, gentle and respectful as always, kindly requested I stay with Erich while he formulated a plan.

Once Erich caught his breath, Rafa directed him to hold onto the stern deck of his kayak while Rafa paddled him downriver to meet the rest of the team. This was a pretty ballsy move since none of us, and only two other people in known history, had ever paddled through this narrowing canyon.

"You sure about this?" Erich asked, having just been saved from drowning.

"You bet! It'll be easy. Just kick your legs and use my kayak like a kickboard." Rafa could sell ice to penguins. That optimism could've had us jumping off cliffs. If I could choose anyone in the world to inform me of my parents' death, it'd be Rafa.

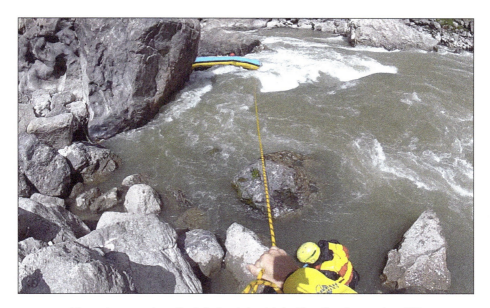

After making it to shore, Erich (yellow helmet) helps Tino (hand on rope) and West attempt to rescue the raft. Photo by Tino Specht

My job was to go ahead of Rafa, stay in the middle of the river and not screw up. I did my best not to become another liability as I shot between the towering canyon walls without my chaperone.

Rafa followed with the bright yellow bow of his kayak sticking high into the air, with Erich clinging to his stern deck. The river careened around the corner and gathered speed as it squeezed between perpendicular walls of rock that blotted out all but a letterbox slit of daylight, high above. Even without paddling, the river carried me along far faster than I was comfortable traveling through the unknown. There were a few small holes to punch, so I kept my paddle brace low to protect my shoulder. The sheer canyon walls meant that I would have to run every rapid blind. If I picked the wrong line, I could get sucked into a hydraulic or trapped in a sieve until hell froze over. My ass was so tight you couldn't pull a banjo string through it.

I whipped past Jeff so fast that I couldn't process his appearance until I was well past him. He was safe on the shore, shivering atop a rocky perch. There was no place for me to eddy out between the canyon walls, and our shouted words drowned in the roar of the rapids. He gave me a thumbs-up and motioned for me to keep going.

Around the next corner I spotted Juanito, who had rushed ahead with Tino to help Simón and John right the raft. That task accomplished, he sent the raft to the first accessible beach downriver, then waited for the rest of us.

I spun into the eddy with Juanito and watched the raft make a safe landing a few hundred yards downriver. Erich and Rafa eddied out across the river from us. Rafa was winded from wrestling his unwieldy kayak through the twisting canyon with Erich hanging off his stern like a 200-pound sea anchor. Juanito hopped onto a boulder and threw a line to Erich to pull him across the river, but the current ripped the yellow rope from Juanito's hands, sending Erich downstream, where he scrambled up the opposite bank, until Juanito paddled over and ferried him back to our side of the river. Simón and I got a huge fire blazing as the rest of the team, minus Jeff and Tino, straggled in for food and warmth.

The Tino Show

"I learned that courage was not the absence of fear, but the triumph over it. The brave man is not he who does not feel afraid, but he who conquers that fear."

Nelson Mandela

Tino Specht hailed from Massachusetts and was, along with the rest of the Tigers, part of a group of elite international whitewater kayakers known as "The Tribe." Somehow, through dumb luck, I had garnered the talents of The Tribe, of which I knew nothing until several days into the expedition. Membership in The Tribe was governed by a strict set of unwritten guidelines that, from what I could tell, admitted only super-talented, ultra-hip young hotshots who could be trusted to uphold a shared code of honor in some of the most dangerous conditions on the planet. Tino specialized in whitewater videography and had a propensity to go shirtless, revealing a physique that belonged on the cover of some cheap bodice-ripper near the checkout stand. As if that weren't enough, his Spanish was fluent and he was genuinely a nice guy and great team member; a real nightmare for those of us who fall further down the Fabio scale.

When the raft first went over, Tino sprang into action and didn't stop to catch his breath until everyone was safe. He quickly roped Erich to shore, handed him off to Rafa, then jumped into his kayak and shot downriver. He paused long enough to determine that Jeff was safe on the bank, then charged ahead to help Juanito drag Simón, John and the raft to shore. Tino then shouldered his kayak several hundred feet straight up a loose scree slope to an old railroad track and followed it upstream at a dead run, carrying his kayak and paddle. He hustled through an old rail tunnel then slid down another cactus-covered gravel slope, launched his kayak into the rushing water and ferried across the river to the clump of rocks where Jeff was preparing his own evacuation on the other side of the river.

Helped by Tino in his kayak, Simón (red) and John (yellow) float with the overturned raft to safety. Photo by Tino Specht

Jeff was barely above hypothermic and still suffering the effects of altitude sickness. He wore only a pair of rain pants and a short-sleeved river shirt, having left his waterproof dry suit with the bank crew. His breath sounded like a stick against a picket fence. Jeff listened to Tino's rescue plan and proposed an alternative. He'd just climb out of the canyon and make his way back to civilization, which, at this point was over a desolate mountain range and countless miles away. Perhaps he'd grab a hot cup of coffee and a cinnamon scone at a Starbucks along the way. If that didn't work, he would just wait to get eaten by wild animals. Either way, he didn't want to get back into the river. Tino repeated his plan to ferry Jeff downriver on the stern of his kayak, as Rafa had done with Erich.

"Yeah, I appreciate the offer," Jeff told Tino, "but I think I'll just go ahead and hike out."

Tino diplomatically told Jeff to get into the water. Jeff tried to negotiate, and Tino politely assured him this was not a negotiation.

"Get in now, man. Let's go."

Jeff slipped into the chilly water and latched onto the kayak. Tino offered words of encouragement as Jeff shivered uncontrollably and gripped the boat with white knuckles. When they rejoined the group, Jeff huddled by the fire, still shivering and praising Tino for his rescue efforts.

I was relieved to have everyone together and safe. I called Barbara to report the incident and our status, then contacted the ground crew to come meet us, carefully noting the coordinates to match up with our transponders. The raft configuration clearly wasn't working. The flip came in a relatively easy rapid, where the huge boulder should have easily been avoided. Without the immediate and well-practiced efforts of Tino, Rafa and Juanito the result would have been disastrous, and the river was only going to become more difficult.

West calling in a status report below Izcuchaca Canyon. Photo by Erich Schlegel

The Tigers huddled together as I met individually with Jeff, John and Erich for their impressions. John and Erich didn't offer much feedback, but Jeff had been a raft guide on many commercial trips, including the Grand Canyon. He said Simón wasn't providing any leadership. Before committing to any rapid, a raft captain typically briefs the crew on the line he plans to take and the hazards to avoid, and if the rapid is big or complex the team pulls ashore to scout and plan. But Simón just drifted into rapids and started hollering orders in a mix of Spanish and broken English. John doesn't understand Spanish. Jeff and Erich understood Simón's orders but fared little better.

Camp below Izcuchaca Canyon, upriver of Tiger Tooth Rapid. Photo by Erich Schlegel

Simón was a world champion rafter from the Chilean raft team but had little or no experience in commercial rafting. With a group of seasoned rafters, not much talk is required. But we didn't have seasoned rafters. We had a bunch of headstrong canoe racers. It was time to make a change.

Within the hour, David and Jennifer climbed down the cliff to join us. David, Jeff, Juanito and I gathered for an executive board meeting. Juanito gently offered up that I was going to have difficulty getting through the more dangerous rapids, since my roll was less than optimal to say the least. He wanted to paddle the raft as a tandem through the harder sections, just him and me. Juanito also suggested the Tigers make a scouting run to Puente Cucharan before he and I attempted it in the raft. This seemed a bit over-cautious but Juanito was the whitewater leader, so I agreed.

I met with the entire team to lay out the new plan, which would leave John, Jeff and Erich on the bank for most of the remaining 500 miles of whitewater.

That was no problem for Jeff, who had already volunteered to sit out until his health recovered, but John didn't like it one bit. He thought he was being kicked off the team. I tried to explain that it was a matter of safety and efficiency, but John wasn't buying it. For all practical purposes John was getting an all-expense paid vacation down the Amazon in exchange for a little support here and there. I understood that he was upset at not being able to paddle the whitewater, but the fact was that he didn't have the necessary whitewater skills. John's mood was set in stone.

A few of us camped out on the river to watch the boats and gear, while a good portion of the group headed back to the hotel in Huancayo.

It was the end of day eight. We had covered 23.21 out of 4,200 miles.

Chapter Six
Mayhem and Mary Jane

Day 11, Izcuchaca

"We had two bags of grass, seventy-five pellets of mescaline, five sheets of high-powered blotter acid, a saltshaker half-full of cocaine, and a whole galaxy of multi-colored uppers, downers, screamers, laughers... Also, a quart of tequila, a quart of rum, a case of beer, a pint of raw ether, and two dozen amyls. Not that we needed all that for the trip, but once you get locked into a serious drug collection, the tendency is to push it as far as you can."

Hunter S. Thompson

Our day started slowly, owing to a road worker who inexplicably parked his car at the gated entrance to the desolate farm where we put in. David, Jennifer and Erich beat the one-man traffic jam, but the Tigers didn't. After about an hour, David drove up to the road to find the entrance blocked and the van full of Tigers waiting on the other side. Gear and Tigers piled into the truck and David drove them down to the river, then came back up and dragged the offending vehicle out of the way with his truck. At this the road worker finally appeared, looking mad enough to fight until Jennifer calmed him with a smile and a couple of cold beers.

After a couple of days scouting and re-arranging the team and boats, Juanito judged this stretch sufficiently safe for Erich to join us in the raft, though he extended no similar invitation to John, and Jeff was still too sick to paddle. So, on the same day the spacecraft Voyager left our solar system, we shoved off with Erich, Juanito and I in the raft and the remaining Tigers in the four kayaks.

Tiger Tooth Rapid. Photo by Erich Schlegel

Within a few hundred yards we came to Tiger's Tooth Rapid. There was no talk of running this monster, which nearly killed Rocky when he and James Duesenberry stumbled upon it at dusk during their Mantaro first descent earlier in the year. Rocky was leading as he swept around a bend and saw the entire river pummeling into a massive hydraulic. With no time to avoid the hole, Rocky squared up and charged into it. The hydraulic flipped Rocky's boat and thrashed it like a pit bull with a new chew toy. Duesenberry, following a few boat lengths behind, saw Rocky disappear into the maw and was able to move left to hit the river-wide hole at its weakest point. Even still, it cartwheeled his boat several times before finally releasing its grip.

Duesenberry later estimated the hole held Rocky for at least a minute, perhaps two. He had time to catch an eddy, get out of his boat and, finding no vantage from which to get a safety rope to Rocky, climb back into his boat and secure the spray skirt for whatever came next. Finally, Rocky kicked free of his boat, which flushed without him. Duesenberry began to chase it through a train of exploding waves. Moments later Rocky flushed from the hole and clawed for shore, his body starved of oxygen and his waterlogged dry suit dragging him down.

At that moment Duesenberry spotted Rocky's helmet from the crest of a wave and swooped in for the rescue. By the time Rocky crawled onto the shore his boat was long gone, along with his camping gear and food, GPS receiver, MacBook Air, video camera and copious river notes. Neither he nor Duesenberry knew what lay downstream, but both knew that if Rocky were caught in another hydraulic he would very likely drown. After a fitful night on the rocks, Rocky crossed the river hand-over-hand on an old cable bridge, climbed 1,000 feet out of the sheer canyon and hitchhiked to Izcuchaca, 22 miles downstream. Duesenberry continued alone through the canyon, where he found Rocky's boat pinned on a rock. Rocky hiked in and continued the descent with Duesenberry from Izcuchaca down. Months later Rocky returned to run the section he'd skipped and claimed a completed descent of the Mantaro.

Now it was our turn. The Tigers are some of the best whitewater paddlers in the world, but today this particular rapid was king. With little discussion, Juanito and Rafa organized a set of ropes to line the unmanned raft through the deadly rapid. Unable to sheath my paternal drive, I lent my girth to the effort by holding onto Rafa to keep him from being pulled into the rapid by the taut rope.

With the raft and kayaks now bobbing safely below Tiger's Tooth, we waited as Erich tinkered with his underwater camera housing. Now that we were below the most dangerous rapid, Tino and I switched, so I could kayak with Rafa. I found a perch atop a boulder that gave me a fine view of the river. Soon I spotted a shiny bright greenish bundle about the size of a couch pillow spinning in the eddy. I'd never seen such a thing in person but I had on the local TV news, usually surrounded by a herd of stone-faced DEA agents.

Simón and Rafa scout Tiger Tooth Rapid. Photo by Tino Specht

I sliced open the couch pillow with my river knife. Yep, it was pot. Reefer, grass, weed, Mary-Jane, or shit—whatever you call it there was about eight pounds of the stuff, densely packed, watertight and ready for unbridled capitalism. Immediately, I scanned the cliffs for the owners of this surprise bounty, but didn't see a soul.

Rising from my squat, I climbed over to Simón and Rafa who were sitting next to the raft in their kayaks. Instantly my status changed from "dude we have to babysit" to MISTER POPULAR! Simón was happier than a wiener dog on stilts. He was speechless, but quickly devolved into whoops and hollers. Tino, Erich and Juanito scampered over to see what was causing the ruckus. They passed around the bundle to verify the contents, then all joined in chanting my name over and over at the top of their lungs. Simón, who appeared the most excited, became the bearer of the bundle as we headed down river. As we made our way to Puente Cucharan, we came upon another bobbing bundle of bohemian bounty, just like the first. Cheers and amazement erupted once again. As we came upon another, then another, then another, the cheers diminished and a hushed mood settled over the team.

Even without partaking, the team went from super-stoked to pure paranoia in two hours. All told, we found 14 bricks totaling 25 to 30 kilos and worth about $30,000. Having made it all the way through college without so much as a whiff of pot in my lungs, I was suddenly in possession of a trunk load of the stuff, smack dab in the middle of a heavily militarized corner of the Andes known to be a favorite haunt for drug traffickers. To add to the drama, my wife and teenage daughter were waiting for us at the bridge, and I was leading an expedition sponsored by the world renowned and rather conservative National Geographic Society.

Puente Cucharan. Photo by Tino Specht

I was also acutely aware that on this expedition the buck stopped with me. I called the team together. We agreed that if confronted by any form of authority in uniform we would immediately tell them what we found and hand it over. If we came upon *narcotraficantes* looking for their lost goods, once again we'd hand it over right away. For anyone else we'd just smile and wave. No one on the shore crew was to know what we'd found, save for David who as team manager had to know everything.

After Tiger's Tooth, Tino took to the raft while I got his kayak. Steep narrow canyons gave way to more manageable terrain, allowing the sun to light a scene straight out of any river-runner's bucket list. We cruised through hours of fun water riddled with easy drops, non-lethal holes and dozens of play waves. Rafa and I snacked on chocolate bars and salami, and he taught me some tricks that didn't exist back when I paddled whitewater.

My shoulder was holding up with the help of half a hydrocodone every few hours. Every turn in the river brought another wonderful surprise: picturesque villages, landslides with house-sized boulders and small farms with orchards. Hot springs dripped from tall cliffs, leaving cave-like stalactites. We drifted under them and let the warm water wash over us.

Now and then a Tiger would come up with another brick of grass, but the celebrations dissipated into quiet apprehension.

Juanito with our cave-full of wacky-weed. Photo by Tino Specht

A few miles before Izcuchaca, the canyon widened to meet the paved road from Huancayo, which we paralleled. We found a cave across from an old hot springs resort. The English traveler Ambrose Petrocokino made note of these same hot springs during his trek from Ayacucho to Jauja in 1902. *"We passed a peculiar formation of rock on a high bank opposite, and then a large spring depositing matter, like a dropping well."*

We formed a dunnage chain to unload the cargo up a slope into a cave, for safe keeping. All that was left of the contraband was an odiferous green slurry sloshing around the bottom of the raft and Simón's kayak. It smelled like the flavorless soup you get in a vegetarian café.

Tino, Jennifer and Rafa catch a mototaxi into Izcuchaca. Photo by Erich Schlegel

Chapter Seven
The Rocky Road Less Traveled

"When your dream is your primary focus, there will be casualties."
                                        Ethan Hawke

Izcuchaca Bridge. Photo by Erich Schlegel

Persistent rain overnight raised the river level a couple of feet. Under the colonial bridge leading from the main road to Izcuchaca, the Mantaro squeezed through a narrow gap and rushed headlong into loud rapids that were mere ripples the day before. The Tigers, not a bright-eyed morning bunch, slowed to a snoring stumble as they made efforts to start the day. After several cups of hot coffee, we gathered for a planning session in one of the closet-sized diners around the plaza. The night before, David managed to get an anemic online connection for a few minutes to download the expedition email. One stood out from the bunch:

Hey West,

I see from your blog posts that you're probably near Izcuchaca. I can possibly come join you now. How long do you think it will take you to walk the 92 km downstream of P. Tablachaca? Will you be carrying/dragging your boat the whole way?

You've done a pretty good job keeping the Mantaro under the radar, though I did notice a few pictures and text that give the location away. You might want to be a bit more careful about what is posted until you get to the end.

Rocky

Not having seen the Mantaro River downstream of Tablachaca Dam during the summer, Rocky assumed the section was dry, and therefore impossible to paddle. But before we started down the Mantaro a week earlier, Piotr led me and other team members on a scouting trip to assess the water level below the dam. To our surprise, we found the flow more than sufficient to float a kayak.

Rocky didn't know that, and his email struck me as a not-so-subtle effort to insert an asterisk into my source-to-sea expedition. He was laying the groundwork for the argument that I couldn't have paddled the entire Amazon because there was no water in the 57-mile stretch below Tablachaca Dam. I found the implied threat annoying but empty because I knew of the presence of water in Tablachaca Canyon.

Dancing on the Izcuchaca Plaza de Armas. Photo by Erich Schlegel

The dance posse. Photo by Erich Schlegel

Tiny dancer in Izcuchaca. Photo by Erich Schlegel

The only portion of Rocky's message that caused us any real concern was the thought of him joining the team. His relationship with the expedition began to fracture in the months prior to the launch, when he said I should represent myself as part of his expedition—rather than the other way around—in any correspondence involving the subject of the new source. Back in April, just days after giving me his permission to start my expedition from the new source of the Amazon, he sent me this gem:

> [T]here is not as much of a need for you to worry about claim jumpers to run the Amazon source-to-sea. I think the only person you really need to worry about is me, and I do still consider myself to have the right to be first in that regard, as it does not affect your original goal to simply be fastest down the Amazon. I'd like to hear your opinion of whether or not you agree with that, so please write back.

I didn't respond.

This was the first indication that Rocky believed that he could retract his permission for me to start from the new source. My view is that once I agreed to his requests and he'd given me his blessing to start on the Mantaro, there was no turning back. At that point, I clearly and repeatedly reported to him and National Geographic that my primary goal was no longer a solo and speed descent, but the much more notable first complete descent of the Amazon from its most distant source.

Rocky's goal, as I understood it, was to run the Amazon's five main Andean tributaries and substantiate his new source claim with boots-on-the-ground GPS measurements. In the process, he had already participated in the first descent of the Mantaro, though he still had yet to complete the section from Tiger Tooth Rapid to Izcuchaca. He wrote that he wasn't interested in paddling 3,800 miles of flat water to claim the first descent of the entire Amazon system, though he certainly recognized the significance of that achievement. We had a deal, set in writing with no stipulation for retraction.

Rocky and all the members of my Amazon Express expedition team agreed to maintain the "fastest" goal as a mere façade, in order to keep the new source under wraps. National Geographic maintained the exclusive right to reveal the new source, which they effectively purchased with their sponsorship of Rocky's research and my expedition. The magazine planned to publish an article including both expeditions after my trip was complete, and after Rocky had published his source discovery in a peer-reviewed geography journal.

Rocky reasonably thought the National Geographic story would be about him and his discovery of the new source. The editors had something else in mind, and when Rocky finally pieced together their plans he was none too happy. Until that point, Piotr handled most communications with National Geographic on Rocky's behalf.

When Rocky discovered National Geographic's plan for an article featuring both of our expeditions, he launched a barrage of emails that raised some eyebrows at National Geographic headquarters. Rocky's messages were full of capital letters, exclamation points and implicit hostility, according to one staff member. They were far from the refined and somewhat stifled maliciousness inherent in the upper echelons of oak-paneled boardrooms with which the folks at National Geographic were accustomed. Rocky managed to get their upper lips in a curl.

Rocky got under my skin too when some of my sponsors and members of my team said he'd sent them disparaging emails about me, as well as to National Geographic. It was one thing when his anger was directed only at me, but now his venom was affecting the success of the expedition. At no time did I negate the impact his discovery of the headwaters or his truly gracious support for our first descent, but now, as he held the threat of sabotaging that support over my head, my patience was all but played out.

By asserting that he had a "right to be first," Rocky's email posed a real and direct threat to the principle goal of our expedition. Piotr and David were equally stunned, and we all agreed I should handle Rocky with kid gloves. Because I could think of no way to respond to Rocky without riling him, I simply didn't respond. As his messages became more vitriolic, I continued to ignore all but the most non-confrontational ones. So long as we were making forward progress downriver, anything that wasn't physically in our way wasn't a priority, and this included Rocky.

Given Rocky's volatility, we were taking a chance that he might decide to start paddling down the Amazon in a bid to claim the first descent for himself. He already completed most of the Mantaro so he had a sizable head start. He made sure I knew it, too. "I am your only competition for the first descent," he told me four times during a video chat before I left Texas.

For weeks I rode a razor's edge between provoking Rocky to paddle the entire Amazon himself—with a 500-mile head start—and acquiescing to his unreasonable demands. This left me in a constant state of anxiety as we raced against the clock to complete the Mantaro. Every day that Rocky didn't start down the flat water was another day we had to catch him if he tried. We calculated our potential speed as experienced ultra-marathon racers in fast sea kayaks versus his speed in pretty much anything he could get his hands on. Sea kayaks weren't available in Peru, so we figured he would most likely use a whitewater kayak or dugout canoe.

Rocky's email arrived at just the wrong time. I was in no mood to put up with his nonsense, and his lead had diminished just enough that I didn't have to. With 20 years of racing experience under my belt and 3,800 miles of track ahead of us, this was my goddamned football. Even if he started on the flat water tomorrow, I could catch him.

It was time to cut ties.

Huddled with Jeff, David and Juanito in that tiny café in Izcuchaca we formed a plan on how to respond. I wanted the leadership team to weigh in. Jeff was solid in his opposition to Rocky joining the expedition. David was more measured, balancing the pros and cons, but concluded that Rocky would be more of a hindrance than a benefit on the river. Juanito saw no advantage, at all, in bringing Rocky on board. His focus was the safety of the team and he had no interest in including an unknown entity.

In his day job as a corporate big shot, David had mastered the polite-but-unequivocal language of high-stakes firings. He drafted the email, then ran it by Piotr and the rest of us for approval:

Dear Rocky,

Thank you for your interest in connecting up with the Amazon Express team as it continues downriver. Given the tenor and content of the feelings expressed by you about West and the Amazon Express, to National Geographic and others involved with the expedition, we feel it best to continue without your participation on the team.

We will continue to credit and support your research and findings related to the Río Mantaro as the longest source of the Amazon, and do our best to keep the information contained until the appropriate time.

David Kelly, Amazon Express Whitewater Team Manager

David sent the email, cc'ing Piotr, and we turned our attention to the river. After about four easy hours and 27 miles we reached the reservoir above Tablachaca Dam. Juanito and I agreed that this specific run was a perfect commercial raft trip—manageable, but exciting enough for novices with professionals at the helm. With time to chat between read-and-run Class IV rapids, he and I discussed the politics of the expedition and what remained.

At one point during the easy day, he asked, "Why does Rocky want to join the expedition at this point? It doesn't seem as if he likes you very much."

I took the question at face value, "Initially, I invited him before I knew him very well. Now, I think he believes that he has a right to be on the team, due to offering up his discovery of the new headwaters and the promise I made to seek his approval for changing where we started the expedition."

"But, you are paying for the expedition," he replied.

"Yeah, it's all very strange to me, too."

Juanito and West, Izcuchaca to Tablachaca. Photo by Erich Schlegel

Chapter Eight
The Vegan Adventure Tuber

"They say that God watches out for children, drunks and fools. Simply put, this means I stand a two-thirds better chance for divine intervention in my life than most other people."

<u>Lenny Castellaneta</u>

After our very fast run to Pressa Tablachaca, we drove back to Huancayo to prepare for the middle Mantaro. That night, after an afternoon of napping, organizing and cleaning gear in the hotel courtyard, we all met Lucho at La Cabana for dinner. Filled to the brim with pizza and hot rum tea, we watched videos and reviewed photos from the first two weeks of the expedition, which included all the scouting we accomplished from the roads.

By this time, we had planned to be nearly finished with the whitewater rather than barely a fifth of the way, which explains why flatwater team members Ian Rolls and Jimmy Harvey showed up as we were finishing supper. They were fresh from Texas by way of Lima, where they picked up the sea kayaks and drove all day to meet us in the Andes. Jimmy planned to paddle with us as far as Iquitos, more than 1,000 miles below the Mantaro, where Pete Binion would take over his seat in one of the tandem kayaks. However, we were so far behind schedule that Jimmy never did get to paddle with us, because of the commitments he couldn't change.

Though there only a few days, Ian and Jimmy added to the festive mood. We shared more rum tea and as we passed the laptops around the table Jeff stumbled across a message that stopped him cold in his tracks. The mother of Davey Du Plessis, a young South African on a solo Amazon expedition, had posted a desperate plea to an explorer's website. Apparently, Davey had been shot on the Ucayali River, an Amazon tributary that we would soon be paddling ourselves. His mother was seeking someone, anyone, who could help her get information about her son.

I traded emails with Davey a few months earlier. He wanted to be the first vegan and first soloist to complete the source-to-sea route of the Amazon River, though he didn't know about the new source on the Mantaro. We wished each other well and I monitored his progress as he blogged. Later that evening, from the hotel lobby, I picked up the satellite phone and called South Africa. I barely got out my introduction before Davey's mother broke into tears as I tried to interpret the fastest South African dialect my slow Texas ears had ever heard. I didn't understand most of what she said but got the gist of it. She was clearly distraught, so I let her go on for 10 minutes before interrupting her for the sake of forward progress. The crux was that Davey had been shot and was in a hospital in Pucallpa, but she didn't speak a lick of Spanish, so communication with Davey's physician was impossible.

This poor mom was going through hell on the other side of the world as her son lay wounded in some jungle hospital. I told her we'd figure it all out and call her back.

Within a minute of hanging up with Davey's mom I handed the phone off to Erich, who spoke in brilliant Spanish with Davey's doctor. Davey had been shot three times with a shotgun, but was doing quite well under the circumstances, the doctor reported. A chunk of lead lay near his heart, one lung was collapsed and another piece of shot was lodged near his spine. The doctor removed most of the shot but didn't have the facilities for the more delicate surgery Davey needed. He was to be flown to Lima that very evening for more treatment. The doctor asked us to call back in 30 minutes when Davey would be roused for the flight.

I called Davey's mom with Erich's report with a promise to call her as soon as I spoke with Davey. A wave of relief bounced off the satellite from a world away. Not a dry eye in the room. I told her the name of Davey's receiving hospital in Lima and left her much calmer. Right on schedule, I called the physician's cell phone and he immediately put Davey on the line. He spoke quietly but clearly, and slowly enough for me to follow his story. He was quite lucid and recounted his ordeal in great detail from the initial attack until his arrival in Pucallpa.

Davey told me that he floated down the Ucayali River on a truck tire inner tube. I wasn't sure I heard that correctly, so I asked him to repeat it. Apparently, his kayak had been lost, stolen or destroyed, so being a tough creative type, Davey commandeered one of the inner tubes local people use as fishing rafts. Davey said he'd been floating along the outside of a curve, where the water is faster, when a fellow appeared out of the jungle. Davey smiled, waved and offered salutations. The man ignored him, which is normal in those parts. He disappeared into the woods and emerged a few seconds later with a second man carrying a shotgun. Without word or warning, the gunman raised his weapon and fired, hitting Davey square in the chest. Davey flew off his inner tube. Two more shotgun blasts struck him as he swam for shore.

Davey made his way to dry land and plunged into the jungle, where he collapsed. Once he figured out that he wasn't immediately dying, he staggered off to find help. He came across one or two people who ran off, startled by his bloody appearance. Eventually, he stumbled into a small village where he found a woman and her nephew who said they would only help him if he paid them money. He had no money. It was in the inner tube with everything else he owned.

Eventually, the villagers laid him in a small wooden motorboat, or pecky pecky, and covered him with a plastic tarp while they decided what to do. For hours he lay there, not knowing whether the people meant to help him or if they were hoping he'd just hurry up and die. Finally, a small group came down to motor him back upriver, towards Pucallpa. He was transferred from one pecky pecky to another throughout the night and all the next day, finally reaching the hospital in Pucallpa. As Davey recounted his story with only one working lung, I didn't have the heart to ask any further questions. I was curious whether he'd ended up paying for the ride to the hospital and what he'd been using to propel his inner tube, but neither question was important. (In his book *Choosing to Live*, Davey reports that he was in a kayak and shot by men who approached him in a pecky pecky.)

A calmer, joyous mom answered on my first ring. She was relieved to hear Davey was awake and seemed likely to recover fully. I gave her the telephone numbers of some of our English-speaking contacts in Lima and urged her to call them if she needed anything. We hung up with a promise that I'd check on Davey over the next few days and get back to her, which I did. A few days later I received a text message from Davey's mom, reporting that Davey's father had flown from the U.S. to be with Davey, who was recovering in Lima.

A year later, Darcy Gaechter became the first vegan and first woman to paddle the entire Amazon River; however, the first inner tube descent is still wide open.

Downtime in Huancayo, in contact with DuPlessis and his mom. Photo by Erich Schlegel

Chapter Nine
The Big Bend

"There is no such thing as work-life balance. Everything worth fighting for unbalances your life."

Alain de Botton

With hard plans for an early morning departure from Huancayo to the Río Huarpa confluence, we got away just after lunch. Under the flimsy guise of illness, Simón remained at the hotel to take care of business with plans to join us in a couple of days. Isabella stayed back to get some respite from the adults. There were TV shows to binge watch and friends to email.

Robin Hanbury-Tenison and Richard Mason traveled the road from Huancayo to the Río Huarpa confluence during their 1958 expedition to cross South America in the newly developed Citizens Jeep ("CJ") adapted for civilian use after WWII. Traversing swamps, grassland and mountains, they drove from Arequipa through Ayacucho, then north along the Río Mantaro on this very road. After a night in Huancayo, they continued through La Oroya and down to Lima, where they completed their journey. Given the number of times their jeep was broken-down or mired to its fenders in mud, this road must have seemed quite the breath of fresh air to them, though it rattled our molars and strained our bladders. The caved-in goat path pretending to be a road alongside the river hadn't changed much since then.

David, in the lead van, drove with wild abandon, blasting around dusty hairpin precipices and passing prehistoric trucks in a race against the sun. No one relished the idea of traveling this road in darkness. That consummate middle-child of the Spanish empire, Pizarro, remarked on these Incan and pre-Incan highways with wonder, comparing them favorably to the Great Pyramids of Egypt. Ambrose Petrocokino, during his 1902 expedition, more accurately described this specific road. *"A few loads of granite setts shot into a road and left to settle would fairly describe the present condition of this famous highway."*

On the road from Pressa Tablachaca to Anco. Photo by West Hansen

Late in the afternoon, with about 10 miles to go, we crossed a substantial bridge at the spaghetti western town of Mayocc, where a few bored military personnel waved us through a checkpoint. Mayocc is well represented on the map drawn by Padre Manuel Sobreviela and his confrere, Fray Narciso Girbal y Barcelo, in 1791. Sobreviela was a Franciscan missionary who explored, mapped and journaled the entire region between the Apurímac and Marañón Rivers, including the Mantaro and Ucayali. His hand-drawn map, used 60 years later by Herndon and Gibbon, recorded the headwaters of the Mantaro River at Lago Chinchaycocha, later renamed Lago Junín.

Though this was the dry season, the Mantaro flowed at a healthy clip 50 feet below the modern steel bridge just outside of town. We were fortunate to find the prediction of a dry riverbed far from accurate, with a healthy flow rolling down the Mantaro.

Late in the day our caravan bumped down the rocky embankment to Puente Allcomachay, a single-lane wood plank bridge suspended on steel cables near the mouth of the Río Huarpa. The drivers of dump trucks overloaded with gravel and rock trusted their fate to this improbable structure on their creeping journey to some hidden place in Peru that isn't composed entirely of gravel and rock. Here we unfolded ourselves from the vehicles and began our methodical preparation for the river.

Family Vacation meet Death Race 2000. Photo by Jeff Wueste

Puente Allcomachay over the Rio Huarpa at its confluence with the Rio Mantaro Photo by Tino Specht

After a full day on the road my mind was more attuned to a hammock rather than preparing for whitewater, but we set about inflating the raft, packing drybags and saying our goodbyes to the shore crew. We pushed off at about 5:45, with a couple hours of daylight remaining and the river carrying us north-northeast with the afternoon sun over our shoulders for the first time. Even with a good volume of water we still stuck on gravel bars here and there until the desert canyon closed in and the waterway grew deeper and swifter. Six miles into the canyon, deepening shadows gave us reason to pull over before darkness enveloped us completely. A short day but a productive one, given the grueling 80-mile drive to the put-in at the bottom of the Mantaro's big bend.

Our campsite was near the site of the Kuntur Sinqa landslide, where in 1945 the whole mountainside collapsed into the riverbed. The slide created an earthen dam more than 300 feet tall, blocking the river for 73 days. No one knows how many people were swept to their deaths when the obstruction finally gave way due to the extreme remoteness of the lower river and the poor communications of the day.

Tino found an old campsite on the outside of a sharp right-hand curve, across from a mountain that rose from the river not quite straight up, but damn near. A fire ring in the middle of a mesquite grove indicated that the place had been used on occasion, perhaps by shepherds. Tino and I slept in hammocks, while Juanito bedded down near the fire pit. Rafa spread out his sleeping bag near the boats to keep watch. Before long, the stars came out in droves to fill the narrow gap between the blackness of the canyon walls. I found sleep quickly, under the influence of the hydrocodone that didn't quite quell the pain in my hands, which were cut and swollen from the daily beating they endured against the rocks. The boys stayed up late around the campfire, talking, laughing and passing a bottle of pisco.

Well after the fire had burned down, we were startled awake by a huge rumble heading down the mountainside in the utter blackness. The sudden shock triggered the onset of my stomach maladies, and I stumbled down the bluff to the water's edge with the feeble excuse to check on Rafa's well-being. Rafa shined a light across the river into a cloud of dust hovering over tons of debris that came to rest at the base of the mountain. This would be our only avalanche, though I'd be visited frequently by gastrointestinal issues for the next few thousand miles.

The next day the middle canyons narrowed and split around monstrous boulders. Patterns in the walls and boulders mimicked the opulent draperies and ornate furnishings of a Victorian mansion; multi-colored stripes, polka dots, rectangles and glitter. This section was littered with hot springs (*aguas termales*) and waterfalls (*cascadas*) that flowed from gorgeous bluffs and tall cliffs. For the sake of speed, we resisted the temptation to explore the wondrous terrain, lounge in the hot springs and drink the pure water flowing cold and hot from the limestone cliffs. Under calcified stalactites dripping hot streams of water, we rinsed the dried sweat from our faces. The rapids on this stretch of river to the settlement of Baños Cortis were frequent, though none required scouting or portaging. It was one of the most picturesque sections of river along the Amazon streambed.

Juanito paddling by one of the hundreds of waterfalls spilling into the middle Mantaro. Photo by West Hansen

Tributaries joined the Mantaro frequently, each an unspoiled display of natural beauty with water dropping from pool to pool, until finally spraying down to some pristine gravel bar, where we filled our water bottles. Now and then, an abrupt wind rushed up river, requiring us to actually put in some effort to propel the raft forward. Juanito and I agreed this proved yet another perfect commercially viable raft section.

Tapestry boulders. Photo by Tino Specht

Tino and Rafa had a relaxing time, often drifting with their kayak, skirts off. Tino whipped off pictures with his monster-sized Nikon, kept in a dry bag between his knees. With each bend, we were met with the perfect camping spot, sure that we would never find one so perfect again, yet there it was, around the next curve, more idyllic than the last.

In contrast to the superlative conditions, my gut remained unsettled. I sipped water constantly to stay hydrated and to flush out the source of my discomfort. Lack of an appetite left me weak, though my spirits were pretty good. Juanito and the boys seemed genuinely happy to be on the river, diminishing my concerns that they'd be bored ushering me downstream.

Now and then an ancient cable or rope bridge crossed the river, similar in design but not material to the hammered grass variety made famous in Thornton Wilder's *The Bridge of San Luis Rey*. Rusted steel cable was common, though most of these donkey-width bridges used thin branches for planks. Occasionally we floated under a well-maintained bridge with rough hewn lumber for treads.

As we paddled the canyons grew deeper and narrower, and the daylight thinned early. On we pushed, passing perfect campsites, one after another. With darkness closing in, we donned our headlights and continued threading rapid after rapid. The kayakers faded into the darkness, visible only by the glow of their headlamps. With every new roar of a rapid out of the darkness, Juanito and I braced ourselves and maneuvered that glorified inner tube through whatever obstacles popped into the faint beams emanating from our foreheads as we constantly scanned the blackness. Still we pressed on, determined to meet the team at Sala de Maquinas as appointed, where we were to resupply and take on Erich and Simón.

Finally, we rounded yet another sharp bend and came in full view of the lights of San Pedro de Cordis, high on the outside ridge of a right-hand bend in the river. Crossing under a modern steel and concrete bridge, we scouted the banks for our support crew, though we had no plans for such a meeting. Bright lights, industrial noise and rapids assaulted our senses after the peaceful day. Floodlights dilated my eyes, leaving me night-blind. With the city lights over our left shoulders, we entered another canyon with walls closing in tightly. The rumble of mining machinery was replaced with the roar of water slamming into boulders. Given the noise level, I judged this to be a rapid best run in the light of day. I called Rafa and Tino over to the raft and announced, "I'm more than willing to follow you boys through the gates of hell, but we aren't going to run this rapid tonight. Let's pull over and camp." Thankfully, Juanito didn't object or pull rank.

Though they treated me with utmost respect and friendship, the camaraderie shared by the Tigers was evident in their inside stories and common experiences. It gave me a really healthy perspective on how things must be when someone joins my group of friends. This bond held by the three Tigers made me feel all the better that they had gone out of their way to include me in conversations and ask about my experiences. They truly were some of the most respectful young men with whom I've ever been acquainted, and a credit to their generation.

In the light of morning the rapid was not as loud or fearsome as it had seemed in the night. We ran it cleanly and, in the afternoon, reached a section of river blockaded by huge boulders, exceptionally massive even by the Mantaro's standards. The garishly patterned tapestries of rock climbed straight up from the river. At one point, the gap we had to squeeze through was about three feet narrower than the raft. Thankfully the current, though deep, was slow and steady.

Juanito formed a plan for the two of us to stand on one side of the raft and pull up the other side using a short piece of rope. This half-cocked profile would make the raft just skinny enough to pass through the narrow gap. In typical Juanito style, he adopted an overly paternal tone and explained the plan in about four different ways, talking slowly and deliberately.

Late night campsite below Banos Cortis. Photo by Tino Specht.

We got into position, balancing on the raft's left tube. With the two ropes in hand, we slowly lifted the opposite side of the raft out of the water. In spite of our cautious effort, we suddenly hit the tipping point where the suction of the river's surface against the floor of the raft gave way and the right side shot up into the air, dumping us both into the deep water. I grabbed a line to keep from being swept downstream. No big deal.

Juanito, ever the responsible leader, immediately called out from the other side of the raft, afraid I was being held under or swept down river. I assured him that I was fine. His concern for my safety, while truly gracious, was a bit much. I was inclined to let him know that this wasn't my first rodeo, however I found it best to bite my tongue and let him remain comfortable in his role. It was his duty to be overly patriarchal in his line of work as a raft guide, and it wasn't mine to question. I swam around and climbed the boulder forming the left side of the gap, while he wrestled the raft to its semi-upright position. Working together we squeezed the raft through. Then I jumped in and we were off.

From my perspective, we ran the raft really well. I felt confident enough to take some strokes, either forward or backward, before Juanito called out "front paddle!" or "back paddle!" Whenever I did this, Juanito would stop me and demand I don't make any move without him telling me to do so. While I wasn't nearly as experienced in whitewater, I did have a few decades of river running under my belt and knew how to position the raft. But still, I just let it go. In the big scheme of things, having to wait until Juanito called out an order was in no way slowing down or hindering our progress to the Atlantic Ocean. He shouldered the burden of leading this whitewater section with a relative novice on board, so this was a good time to stow my ego and let him bark his orders.

Preparing to line the raft through a Class V. Photo by Juanito De Ugarte

Several of the rapids were just too dangerous to run, so we lined the raft with long ropes attached to the bow and stern. Fortunately, none of the rapids were blind. All of them had a substantial bank on which we could scout or portage. We tried to run them all. Many of the scouting stops ended in deep discussions about possible routes and dangers, though I was pleasantly surprised to see the three Tigers coming to a consensus on how to approach every rapid. You'd think a rafter or kayaker's most confidence-inspiring quality would be the skill to run extremely dangerous rapids, but my faith in the Tigers was bolstered even more by their willingness to accept that some rapids were just too dangerous even for their extraordinary abilities.

Lizet catching a ride down to Sala la Maquinas. Photo by Erich Schlegel

Most of the tougher rapids were easy to portage. Photo by Tino Specht

Chapter Ten
The Water Cannon

"It's not the heat. It's the stupidity."
Mary Karr

After our usual morning ritual, too late for me and too early for the Tigers, three miles of easy paddling brought us to the Sala de Maquinas, where water diverted from the Tablachaca Dam drives seven massive hydroelectric turbines producing a quarter of Peru's electricity. This diversion is the reason Piotr and Rocky were certain the section of river below the dam would be dry. During the Andean dry season, the majority of the Mantaro's flow is diverted through the tunnel in a man-made quirk of geography that Piotr would later use in his efforts to undermine Rocky's discovery of the Mantaro as the Amazon's most distant source. Piotr's position is that the Mantaro's water is diverted during the dry season and therefore, by the law of selective logic, it is not a continuous waterway and should not be considered the most-distant source.

The trouble with this reasoning is that in the absence of a universally applicable definition of what constitutes a river's course, the question inevitably falls back on subjectivity and politics. When I ran the Mantaro at the height of the dry season there was enough water below the dam to float my kayak, but that point is irrelevant. No one questions the source of the Rio Grande high in the Colorado Rockies, though it runs bone dry for more than 100 miles from El Paso to Presidio. The source of the mighty Colorado River that carved the Grand Canyon hasn't changed, even though the river has reached the sea of Cortez only occasionally during the last 30 years. Moreover, the Volga, Yangtze, Nile, Missouri, Onyx and Murray river are all dammed and have seasonal flow.

In short, when you look at a dry river bed, the river's name hasn't changed, nor can the intermittent aspect of water in the streambed dictate whether the riverbed is or is not a river. If this were the case, then the names and sources of all the world's rivers would change on a daily, weekly or monthly basis. Cartographers, farmers, boundary-setters and nations with river borders would be in constant flux. As odd as it may sound, for the sake of scientific comparison, geopolitical boundaries and common sense, the continuous presence of water does not dictate whether a river is a river. Could you imagine the border between Mexico and the United States changing every time the Rio Grande dries up or floods?

Lunch on the run in the highlands. Photo by Erich Schlegel

The Mantaro's unique topography explains why engineers built the power station here and may also be the reason geographers failed, for so many years, to recognize it as the Amazon's most-distant source. Just below the dam site the Mantaro makes a long horseshoe bend, so that even though the dam and power station are only 12 miles apart as the crow flies, there are 110 river miles and 2,450 vertical feet between them. The loop makes the Mantaro appear shorter on maps than it truly is, and that big drop in altitude creates the tremendous pressure needed to drive the power turbines.

It's an engineering marvel and a spectacular demonstration of the power of moving water—the contents of a good-sized river dropping through a pressurized tunnel and then blasting out the side of a mountain like a 300-foot waterfall turned on its side, with more than 25,000 gallons of water per second arcing almost all the way across the Mantaro. At a little over eight pounds per gallon, water packs quite a wallop. If you think getting smacked with a water balloon hurled by a garden-variety 10-year-old is unpleasant, just consider how it would feel to be hit with a quarter-million of them every second. This is what I vividly imagined as we paddled with no small amount of trepidation toward the water cannon.

Juanito and West under the water cannon. Photo by Erich Schlegel

Juanito's plan was to simply paddle the raft through the narrow gap between the impact zone and the far-right bank of the river. I was apprehensive, but game. As we drew near the turbulent epicenter, a blast of wind and spray stopped us cold. The Niagara-sized stream of water shooting through the air created the world's largest water fan, a hurricane-force headwind which joined forces with the waves and boils emanating from the impact zone to push us into a jumble of boulders. We tried pulling ourselves along the rocks, but the wind and waves made an impenetrable barrier.

We backed off, yelling above the freight train of water. The sound was an unending explosion, a watery volcano that just wouldn't let up. Retreating up river, we made a new plan to paddle under the mountainous arch of water, which in hindsight strikes me as even more stupid than it did at the time. We ferried across to the left bank, from whence the hellish water spout shot out of the canyon wall about 60 feet overhead. Juanito aimed the raft under the arching water and called for full speed ahead.

Sound came from everywhere and shot straight through to our core, shaking my innards through every pore. The river surface trembled and the raft vibrated. Again, the wind slammed us, and before we'd crossed under the spout the current abruptly changed direction. Rather than carrying us downstream, it pushed us violently to the right, toward the liquid vortex where 106 tons of falling water pummeled the surface, sucking anything that strayed close to its perimeter into its foaming maw and straight to hell.

Out of the corner of my eye, squinting against the wind and softball raindrops, I made out Juanito digging hard and yelling something unintelligible against the din. To this day, I'm not exactly sure what he was hollering about, nor did I pause for clarification. ("Pardon, what was that you were uttering?") As the bulbous unwieldy glorified inner tube to which we had entrusted our lives hurtled toward a roaring whitewashed abyss of certain death, Juanito and I were on the same page—something along the lines of *let's get the fuck out of here!*

We turned back upstream and paddled like hell, as errant blasts of water pummeled the tail end of the raft. The water pinged like ball bearings on my helmet and beat bullet shots into my unprotected shoulders. Seconds turned to hours as we fought against the current, barely skirting the worst of the maelstrom. Time for Plan C.

From our vantage point just upriver of the deluge, we paddled once again to the shallows on the left bank, where we climbed out and pulled the raft along. The incessant roar was a lead blanket as Juanito yelled out the plan, six inches from my ear.

"We're going to portage!"

I did my best John Wayne nod to assure him that I approved, as if my approval carried any authority.

"I'm going to climb with the rope!" He screamed, pointing at an uneven jumble of algae-covered boulders under the water cannon.

I stared into his eyes a bit longer before giving the John Wayne nod.

"You stay in the raft and paddle hard!"

This time I gave him a long hard look before forcing John Wayne to the surface.

The thought of paddling the ungainly raft solo against the nightmare we just barely survived left me a tad more anxious than the Duke would have admitted.

If Juanito slipped or lost his grip on the rope I would be shot into the turmoil without a chance of fighting my way out of a place the two of us barely escaped, using all our might.

Juanito unfurled the long bow line and slowly crept 30 feet up the slimy waterlogged rocks that lined the mountainside, underneath the concrete chute. The higher he climbed, the more I noticed the water coming down all around him. Thoughts of what could go wrong cycled quickly through my head, all with a tragic end. Once Juanito reached the end of the rope, I backed the raft away from the shore and paddled hard, using deep strokes against the strong flow sucking me into the morass. The water pounded my helmet and obscured my vision. My goddamned torn shoulder felt like a knife stab on every stroke.

Through squinted eyes I watched Juanito tread gingerly among the slippery rocks, holding the taut rope. Any second, I expected the rope to go slack as Juanito tumbled down the slick mountainside. There was little time to contemplate fear or anything else. Even the smallest amount of energy directed at anything other than the effort to survive was a waste. I swung my paddle hard and deep, not only to ward against drifting backwards into the pit of hell, but to give Juanito relief from any tension on the bowline. We worked together and within a few minutes the line went slack and I was able to paddle the raft into the calm eddy on the downstream edge of the water cannon. Juanito was uncharacteristically quite giddy, describing his shaky-legged traverse and the fear that a slip would be bad news for all involved. We both needed a drink.

West emerging from underneath the water cannon, in desperate need of whiskey and clean underpants. Photo by Juanito De Ugarte

Dynamite Canyon

"Abandon all hope, ye who enter here."
                                    Dante

The Río Upamayo joins the Mantaro at the apex of a broad horseshoe bend, adding impetus to the river as it enters a narrow canyon. The water flowed deep and fast through frequent torrents so we moved efficiently, stopping to scout only the most difficult whitewater. The Tigers leapfrogged through miles of rapids, making a plan for each drop as they paddled headlong into it, a strategy called "read-and-run."

Juanito and I hit our groove. He called out the simplest directions, and I followed them. The Tigers flitted and played in the growing waves, Juanito grumbling now and then about their lack of attention to the safety of our raft, though I felt their attention was enough. We lined the raft through two particularly horrendous rapids, which the kayakers portaged. At this water level, once again, every rapid had plenty of exposed shoreline for scouting or portaging, except one.

Just after meeting the Río Upamayo on the Guittara Curve, the canyon walls closed in tighter than on any part of the river since Izcuchaca Canyon. Normally, this would have caused us to sit up and take notice, but the team's smooth rhythm lulled us into a sense of peace and accomplishment. In direct contrast with the crystal-clear blue skies, distant thunder erupted and reverberated between the high canyon walls. We didn't pay it much mind, continuing to chat and drift along at a good clip. Then a rapid burst of thunderclaps sent shock waves through the gorge and a huge cloud of dust rose from the slivered canyon ahead. Seconds later a shower of boulders rained down on the narrowed river, some as big as bowling balls and others the size of dorm-room refrigerators. Uttering a communal "holy shit!" we scanned the banks for a place to portage but found nothing but steep cliffs and crumbly gravel slides. The river flowed swiftly on as always, carrying us closer to the impact zone.

At the last moment, we all crammed under a narrow overhanging cliff and held onto the kayaks and raft to keep them from being swept downstream.

We knew that construction had begun on the Cerro Del Águila dam project but having passed a large bridge and construction site earlier in the day we figured the location was behind us. Clearly, we were mistaken. The next volley of explosions rained dust and rocks all around us, striking the water like mortars and shot pellets. Our helmets protected us from smaller rocks, but there would be no surviving a direct hit from anything bigger than a grapefruit. As the river squeezed down to 20 feet in width, we dove under an overhang barely big enough for the raft. All around us the river detonated with debris.

None of our options were good. The swift current kept us from paddling back upriver, and the canyon walls were too steep to climb, even if they weren't packed with high explosives. Taking our chances downriver seemed the only option. Having witnessed a large-scale mining operation in New Mexico, I knew that it took hours to set charges and only a few seconds to set them off. Eventually, they'd have to stop blasting long enough to place new charges, giving us a chance to slip through. I tried to time the blasts, but there was no consistent interval between the explosions.

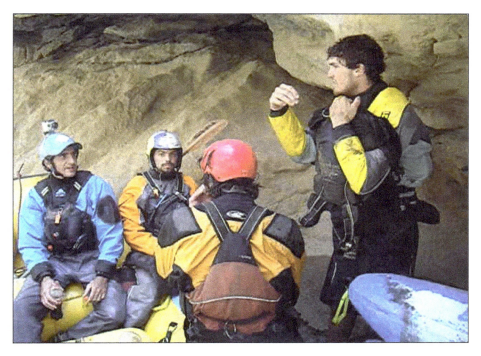

Juanito, Rafa, Simón (back) and Tino discuss a plan to get through the explosions. Photo by West Hansen

I handed my walkie-talkie to Juanito, who crept cautiously into the open to make a line-of-sight transmission, calling out in earnest Spanish, "Please stop dropping rocks on our heads!"

He kept the message simple, repeating it on several channels. Juanito never received a response on the radio, but after about 20 minutes the blasting stopped. We seized the moment and began flying down river. We made it about 100 yards before the river narrowed to nothing and disappeared in a frothing mass of violent waves beneath a massive pile of freshly blasted boulders. Our escape options were a small crevasse in the vertical wall on river right, and a steep slope of powdery pulverized stone on river left, the top of which, several hundred feet up, was the source of the charges. The blasting could resume at any second.

We drove the nose of the raft into the crevasse on river right, as Tino scrambled up ahead of us, holding the bow of his kayak. Rafa and Simón chose the silty sand pile on the other side of the river. I wedged myself into the crack and held tight to the raft's bow line as the current tried to wrench the raft from my grasp. Juanito scampered up the narrow crevasse for a better look. We had to move fast to avoid the next volley. After a couple of minutes, Juanito attached a throw rope to the raft and tossed the other end to Rafa on the opposite bank. Rafa then lined the raft down the left bank while Juanito and I scrambled down the steep right bank toward the pile of boulders that swallowed the river. Tino ferried back across in his kayak to join Rafa and Simón.

Closer to the boulder pile we spotted a wooden construction platform spanning the river over the jumble of boulders. On it stood a man in hardhat and coveralls. He'd heard Juanito on the radio and stopped the blasting so we could get through. We pulled the boats up onto the boulder field along the left bank, took a long lunch break, then portaged over the boulders and continued toward the Atlantic, all the wiser.

The Mantaro disappears into a hellish hole under blasted boulders at the Aquila Dam site. Photo by West Hansen

The Cerro Del Áquila Dam is the last of four large hydroelectric dams planned for the Mantaro, after the Upamayo, Malpaso and Tablachaca dams. When the project was first proposed in the 1970s, engineers wanted to place the dam farther downstream, near the confluence with the Apurímac River. They changed their minds after the great Mayunmarca landslide of 1974, when in the space of three minutes an entire mountainside slid into the river, obliterating the village of Mayunmarca. The impact of the slide striking the valley floor launched debris 200 meters up the opposing mountainside, which dislodged even more terrain. The debris filled the canyon and blocked the river for six weeks, creating a lake 30 miles long and 700 feet deep. When the natural dam finally gave way, several days later under the power of the backed-up river, a wall of water and debris swept downstream, smashing roads, bridges and entire towns. The slide and subsequent flood killed 451 people and pointed out a profound engineering challenge. The Mantaro had shown itself to be prone to massive landslides—the Kuntur Sinqa slide had blockaded the river less than three decades earlier—and the ensuing floods of water and debris posed a threat to any dam.

Engineers wisely planned for the worst, and using data collected from the Mayunmarca surge they designed a dam able to withstand such a colossal hit. To construct such a dam, they had to go well upstream where the canyons were narrower and the rock strata stronger. They added a set of bypass tunnels and a system designed to flush silt from the reservoir. The subject of dams is emotional and messy. Given my druthers, no dam would exist in the world. There are many bad things that come with every dam, often including the flooding of land on which people have lived for generations. There's no way to adequately compensate people for the physical, financial and emotional losses brought about by a dam that inundates their farms, homes and the burial sites of their ancestors. Still, people want electricity and no one has come up with a cheaper or cleaner way to produce it. Water power is perhaps the cleanest source of renewable energy. Peru is doing its best to catch up with countries that began building hydroelectric dams a century ago. In the U.S. and Europe, people who take electricity for granted sign petitions and rally against dams, always with noble intent. But yelling "stop!" without offering a better option rarely yields lasting results. Sadly, no one has presented the Peruvian government with a cleaner and cheaper alternative to damming some of the most beautiful rivers on the planet.

Chapter Eleven
The Middle Canyons

"Do not meddle in the affairs of dragons, for you are crunchy and taste good with ketchup."

Unknown

With the exception of Dynamite Canyon, the three-day stretch from Sala de Maquina to Chyquia Bridge brought some of the best and most technical whitewater of the entire trip, and everyone was in a good mood, especially me. Far from the commercially viable sections of the Mantaro, this section would best be left to whitewater experts. My stomach issues were finally under control and I had much more energy. Prior to this it was the Tigers who were jumping about collecting firewood, cooking dinner and socializing after a long day of paddling. Now, the roles were reversed.

Simón brought along some samples from our Izcuchaca river harvest, which had the Tigers disappearing into the bush for a few minutes at a time and returning quite mellow. Like Rafa, I didn't indulge, though my inventory of aches and pains surely would have warranted a prescription for the stuff from any half-baked doctor in Venice Beach.

Proper positioning in a raft required me to sit sidesaddle on the inflated gunwale, wedge my outside foot underneath the inflated thwart in front and jam my other foot under the thwart behind—a position that only could be achieved by turning my inside foot 90 degrees from its natural position. This guaranteed that my feet would detach from my body long before they slipped out of position. With my feet firmly anchored and my body tossed like a rag doll, my knee joints pivoted not only front to back, but from the newly adapted capability of side to side as well. After a few days of this abuse I was shuffling through camp like a drunk in an earthquake. The upside was that my painfully cracked and swollen hands were no longer the focus of my attention.

Several rope bridges cross the middle Mantaro River. Photo by West Hansen

Simón in some sick brown. Photo by Tino Specht

Being a strong proponent of better living through chemistry, I was always on the hunt for something that would relieve the pain, yet still leaving me somewhat coherent. My tolerance for hydrocodone was low, so I tried to avoid it. I downed fistfuls of ibuprofen, or "Vitamin I" as Jeff referred to it, swallowing the little red pills with the daily dose of caffeine required to grease the morning hinges. The ibuprofen only softened the edges of my shoulder pain, which I still hadn't revealed to anyone but David. Cannabis was out of the question, as I never developed an interest. This left the pisco.

Ah yes, the pisco. While I ranked the local fermentation below a good single-malt whiskey, it certainly was a welcome addition to my evening regimen. A few sips altered my state just enough to sleep, without the threat of a hangover.

The middle Mantaro primarily contains Class II/III rock gardens and rapids. Photo by West Hansen

The Tigers hit the sack early, as my aberrant energy level kept me stoking the fire against an evening breeze rushing down the canyon. I hugged up to the fire nursing the bottle of pisco as wondrous and oddly positioned constellations filled the sky. In modern parlance we refer to these moments as being "mindful," which is to say we're acutely aware of the moment and all it brings. As I sat under that amazing night sky, next to that campfire in a canyon on the Mantaro River in central Peru, free of incandescent light for a hundred miles, my campmates fast asleep and an appropriate amount of pisco under my belt, I recognized that I was experiencing a truly great moment. I absorbed it all for what it was worth, until I noticed the yellow reflector eyes of a mountain lion slowly advancing toward me.

Simón works a boulder field. Photo by Tino Specht

The mindfulness thing didn't change; in fact, the presence of the big cat made me even more aware of the moment. Let there be no doubt, I was pretty goddamned aware. Strictly for medicinal purposes, I took a few more sips of pisco as I stared at the two eyes silently watching me through the darkness. The eyes were about 50 yards off when I first noticed them. I'd seen mountain lions in Texas, and once came face-to-face with one in Big Bend National Park, so I had little doubt these were the eyes of a mountain lion, referred to in the Andes as a puma.

The eyes slowly disappeared behind Volkswagen-sized boulders, only to reappear on the other side. After about 20 minutes they'd floated two or three boulders closer, and I figured notifying the sleeping Tigers would be the polite thing to do. I stood up on my sore knees and stumbled over the bowling-ball river stones to the patch of sand where the youngsters were laid out in their sleeping bags like loaves of bread, fresh from the oven. Tino's hammock was hanging a few yards off in a grove of spiny trees, like a piñata. With as smooth and easy a tone as I could muster I told the boys that a mountain lion was headed toward them. I figured to keep the emotion level a tad understated, letting the words speak for themselves.

"A mountain lion is creeping up on the campfire."

No one stirred, so I spoke in a normal tone to emphasize the point.

"I think we're being stalked by a large mountain lion." It came out like I was asking for the salt shaker.

Still, the dire words elicited no reaction. Granted, the boys were somewhat chemically comatose, so I didn't fault their lack of response. I nudged Juanito's foot and bent over a bit; my knees bitching for more pisco.

"Juanito, we've got a mountain lion about to make a visit. I just figured you'd want to know."

His head popped up a bit, but his eyes remained closed. "How close is it?"

"Thirty meters." I silently congratulated myself for converting to metric, though I may have been off by one or two hundred yards.

Juanito's only response was to lay his head back down.

"I'll notify y'all if it gets any closer."

"Okay… zzzzzzzzzzz." He was petrified with fear.

My moral obligations fulfilled, I retrieved my machete and hobbled back to the campfire. I took my guard duty seriously; after all, someone had to keep a close eye on that bottle. It was difficult to stare into the darkness, so my eyes kept drifting up to the magnificent heavens, trying desperately and wondrously to make sense out of the strange skies of the southern hemisphere.

An hour later, I had yet to be mauled, so I left the last sip of pisco for an emergency and shuffled off to my hammock, machete in hand. For a few seconds, I considered hanging a chunk of salami on Tino's hammock in case he got hungry late at night.

In the morning the cat was gone and it didn't look as if anyone had been mangled. We turned our attention to the remaining distance to Puente Chyquia, which was full of highly technical rapids. We paddled through dozens of read-and-run Class III-IV drops, and the occasional Class V that required extensive scouting and pondering, with discussions in Spanish and English. Eventually, Juanito would turn to me and launch into the plan: "Okay, this is what we're going to do…" Without fail, we hit the exact route he laid out in his instructions, *every single time*. The man was a freaking surgeon on the water.

West and Juanito into the maw. Photo by Tino Specht

Rafa and Tino navigating one of the tight, serene canyons. Photo by West Hansen

Juanito, while highly temperamental and somewhat irritable in his demeanor, was impeccable in his actions. Now and then, a Tiger would sidle up to me and quietly apologize for Juanito's condescending deportment. I'd reply that I was perfectly fine with him, though I don't think they believed me. Perhaps they were under the delusion that I knew what I was doing.

After every big rapid all the Tigers would pronounce excitedly, "Dude! That was sick!" By all accounts each of the rapids was quite sick. Several rapids proved much sicker than others, as evidenced by the emphasis and volume placed upon the word, with a run-of-the mill deadly rapid meriting only a subdued, "Yeah, it was sick."

For me, they were all quite sick. Sick like the bubonic plague. However, by virtue of my generation and paternal status, I am forbidden to use the word "sick" unless I am referring to actual illness or disease. If I did it would be the kiss of death for the word and another adjective would have to be developed at the next Hipper-Than-Thou conference, and that would be a drag.

Just about lunch time, on what we figured was our final day before Chyquia Bridge, we came upon a roaring monster from hell, so we pulled up to the mossy bank on river right and scrambled through the scraggly trees to take a look. Rivulets of sparkling water kept the steep ground moist and muddy across the 100-yard portage. While Juanito stared at the maw of the rapid, we took an extended lunch and nap break. Eventually, after conferring with Rafa and Tino, Juanito announced that we would have to portage the raft as well as the kayaks. The torrent was far too dangerous to line the raft. I nodded, as if my opinion carried any weight at all.

Juanito scouting a Class V+ rapid. Photo by West Hansen

Methodically, we untied all the gear and carried it piece-by-piece up the steep bank, along the rough ground, across the rocky terrain and handed it down a tight 40-foot crevasse to the edge of the river. Then came the raft, which we all dreaded. With every ounce of energy, even from the sweat-allergic Simón, we pulled, heaved and carried the inflated behemoth overland, careful not to drag or snag it for fear of a puncture. The portage ate up a good two hours and far more energy than we truly had to spend.

Tino running it like a pro. Photo by Juanito de Ugarte

We didn't mess around the rest of the day, running rapids quickly and doing our best to make the bridge before nightfall. Our normal banter was reduced to a few pragmatic words. The day was gorgeous and the setting was beautiful, but our thoughts were consumed with plans for the coming days.

Now and then, we spotted a house or lonely electrical pole way up the mountainside, indicating that life beyond this great wilderness was encroaching. Our time was no longer dictated by the rising and setting of the sun, but rather flight plans, long drives and connections to be made. And yet, we still had some considerable rapids and miles of river to get through before returning to the melee of modernity. It was quite the paradox—having to traverse dangerous rapids in desolate canyons in order to climb into a minivan and rush to catch flights in gleaming airports riddled with fast food joints and latte bars. It was all too far and all too close.

On we continued, pushing our limits in a futile race against the setting sun. As darkness finally overtook us, we came to a bridge that wasn't marked on our 30-year-old map. In the darkness we couldn't reconcile the lines of our topographical maps with the mountains surrounding us. I got David on the sat phone, who plotted our location via the transponder and told us we were still 10 miles from the Chyquia bridge, where the support team was waiting. We kept on, moving as fast as we dared through the night.

West pretending he knows what he's looking at. Photo by Tino Specht

About an hour after dark we came upon a particularly loud rapid. Rocky's only notation at this spot was the word "portage." Given his penchant for running things he perhaps shouldn't, I tended to pay attention to any rapid he chose to walk around. We dragged up the shore and half-heartedly scouted the rapid. On the opposite bank was a small house and a dog that barked incessantly. Tired and irritated, I had the urge to swim the river, climb the bank and strangle the dog. In the darkness it was impossible to properly scout the rapid so we pitched camp on rocky ground surrounded by huge boulders. On river left, a light shone in the lone house, where the dog barked for hours.

In the light of morning the homeowners watched as we packed our gear and scouted the rapid, staring blankly despite all my efforts smiling and waving. The rapid had two significant drops that would have been devastating to blunder into at night. Tino, who was bitten with a bad stomach bug overnight and looked even worse than I felt, assured us that he could kayak. I offered up my seat in the raft, but he politely declined. Once he was pried into his kayak, I watched with throw-rope in hand as he, Rafa and Simón ran the drops cleanly like the experts they are. Juanito and I then lined the raft without incident. Our spirits were lighter after sleeping off our heavy fatigue and we knew our ground crew was only a few kilometers away.

We reached the Chyquia bridge in about an hour, just in time for lunch and not too early for cold beer. The desert scrub was a comforting duplicate of west Texas, though the high winds whipping through the canyon were not nearly so welcoming. Finally, we piled into the van and began the slow crawl over the Andes back to Huancayo.

It was good to see Ian and Jimmy, who came to meet us at the bridge along with David, Jeff and Erich, though I was disheartened to hear my fears confirmed that Jimmy would have to leave the expedition. We were simply too far behind schedule. Ian planned to fly back to Texas to spend the next couple of weeks with his daughter before rejoining the team in Ayacucho for the 3,800-mile flatwater leg. Jimmy's departure left a big hole in the team. He was supposed to fill a seat on one of the tandem kayaks down to Iquitos, then switch out with Pete Binion, who was to take the seat to the Atlantic Ocean.

The Tigers, too, were out of time. They would fly out of Lima the next day, as would David. The Tigers had already extended their commitment and simply couldn't afford to stay on the expedition any longer, though they all wanted to.

During the drive, Erich, Juanito and I talked business. Erich's assignment with National Geographic came with an expense account and little oversight. His job was simply to bring back images worthy of the magazine's high standards. How he accomplished this task was up to him. Due to Juanito's concern for safety, Erich had already been forced to skip some of the most spectacular whitewater, and without raft support he would miss the rest of the Mantaro. This was Juanito's chance to shuffle. For $3,000 from Erich's National Geographic budget, he offered to gather a new team of whitewater experts, pay their food, lodging and transportation, and buy an oar rig for the raft.

Until now we had been running as a paddle raft, with Juanito on one side with what is essentially an oversized canoe paddle, and me on the other side with a paddle of my own. With more people in the raft—such as the earlier disaster with Simón and the boys—it worked the same way, with everyone paddling and one guide calling the shots. No guide, not even one as skilled as Juanito, can run a paddle raft through tough whitewater without the active participation of other paddlers. But with an oar rig Juanito would sit in the center of the raft with a pair of powerful levers at his sole disposal, allowing him to spin the craft quickly and move it forward or backward without assistance. The setup would allow Erich to join us for all of the remaining whitewater and he wouldn't have to worry about paddling. He would be free to concentrate on his photography.

Erich getting the perfect shot near Puente Chyquia. Photo by David Kelly

Erich and Juanito shook on the deal. I'd spend the next few days paddling solo on the relatively flat sections just below the source, giving Juanito time to gather his crew before we tackled the rest of the big water as a team.

The precipitous dirt track out of the canyon was quickly lost in darkness, which was fine by me. Jeff drove the van cautiously while David, in the faster truck, stopped at each village to wait for us. The paddlers dozed in the quiet darkness.

The Mantaro, looking upriver from Puente Chyquia. Photo by Erich Schlegel

The road from Puente Chyquia up to San Antonio. Photo by Erich Schlegel

Well after dark, we stopped at a *tienda* where a woman and her two cousins tended a large white plastic barrel of *aguardiente*. Being severely dehydrated and in thin air, I offered restraint, only taking in a polite sip or two. The burning corn liquid landed in my gut with the promise of a quick and torrential hangover, the kind that didn't wait for the next morning but rushed in as soon as the warmth of raw alcohol waned.

Erich had no need for such demure behavior. Flirting easily with the smiling Quechua woman behind the counter, he slugged down several cups of the firewater without pause. The cold gave the raw drink a certain appeal, though the risk of blindness and death dulled my enthusiasm. Ears of corn and hay piles provided a rural familiarity to those of us hailing from similar environs, but with miles to go before we could rest, we pried Erich away from the admiring shopkeeper and drove into the frigid dusty darkness.

Proprietress and firewater. Photo by Erich Schlegel

Crossing the skeletal bridges over the deepening gorge that carved through the canyon, I noticed many of the wooden slats that formed the roadway of a particular bridge had gone missing since we'd scouted this area weeks before. Soon we came to another bridge that had been derelict when we last passed this way, and found it clad with new timber. Each crossing was an exercise in patience and nail biting. All but the drivers, Jeff and David, emptied out of the vehicles, then everyone would shout directions as they tried to balance the tires along the single or double logs that made up the bridge. After a minute or two, the pilots would scream out the window for just one person to give him directions. Then no one would say a thing until the driver called on someone by name to guide him. It was particularly harrowing to see large trucks full of people or crops crossing the same rickety planks.

Jeff, John and West provide expert guidance for the vehicles. Photo by Isabella Hansen

Farewell, Tigers. Thanks for saving my ass. Photo by Erich Schlegel

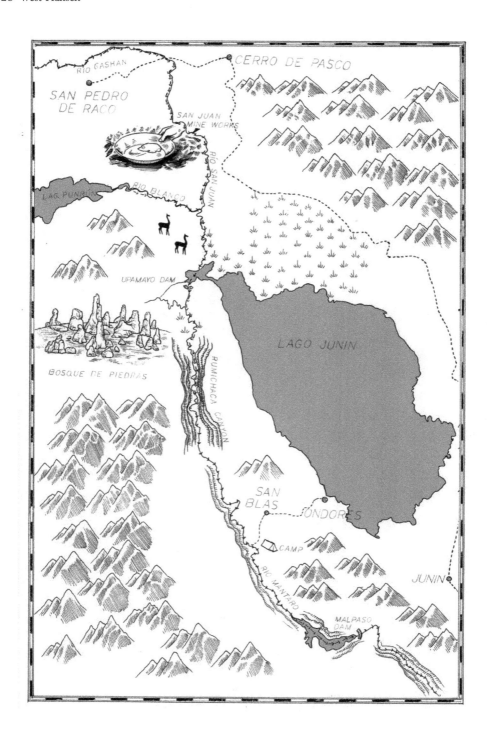

Chapter Twelve
Thin Air and Dirty Water

"About 15 miles from Junín we passed the village of Carhuamayo. Here I saw the only really pretty face I have met within the Sierra and bought a glass of pisco from it."

William Lewis Herndon

Rolling into the altiplano just north of Lago, Junín was like coming home after a long absence. Though it had been only 14 days since Jennifer and David plucked me off the shallow and freezing Río Gashán, this familiar pastureland now seemed a lifetime of dynamite, team drama, water cannons and treacherous rapids in the past. With all but a handful of us flung to the corners of the world, the team was temporarily down to just four.

It was the 5th of September, day 20 of the expedition. According to my original timeline we should have been finishing the 500 miles of the Mantaro, but I still had to paddle nearly 170 miles from the source to Huancayo, then another 57 miles between Presa Tablachaca and the Río Huarpa before reuniting with Juanito and the new Tigers to run the stoutest whitewater yet, the Lower Canyons from Chyquia Bridge to the river's confluence with the Apurímac.

Without much discussion, because of health and skill issues we all knew Jeff and John wouldn't be able to paddle the Gashán, San Juan and Mantaro rivers to Huancayo, or the stretch below Presa Tablachaca, so it was up to me to fill in the gaps in our route.

West taking a break from the Río Gashán in one of the many ruins. Photo by Erich Schlegel

By now, I had the chance to acclimate and knock a little rust off my long-dormant whitewater skills. Feeling substantially stronger than I had on the Río Gashán two weeks earlier, I was able to make the Upamayo Dam in two easy days. The pristine Río Gashán becomes a sewage trough below the town of Cerro de Pasco at its confluence with the Río San Juan. Choking back gags, I shoved myself through rafts of household trash and disposable diapers and met Jeff, Erich and John just below the San Juan mine. Instead of camping on the barren rock beside the roaring mine works, we drove up to the city of Huallyay and found lodging in a hostel on the town square. Though unheated against the freezing night, the dismal accommodations had plenty of heavy wool blankets to pile on. We stocked up on bottled water, rum, bread and cheese before driving back down to the river the next morning.

West strolling down the Rio San Juan. Photo by Erich Schlegel

The San Juan mine. Photo by Erich Schlegel

Colonial church in San Pedro de Pari near the confluence of the Rio San Juan with Lago Junin. Photo by Erich Schlegel

The Upamayo Dam, where the name "Mantaro" begins for the Amazon flow, held back most of the disgusting refuse, and after an easy portage I floated through the otherworldly beauty of the Bosque de Piedras region. The innumerable rocky spires cover the western ridge of the Andes for several miles along the banks of the river. All told, the official natural area covers more than 16,000 acres. The unique spires and other odd formations appear as trees from a distance, hence the name. Rumors of extraterrestrial alien encounters brought more tourists to the area recently, though it's still sparsely visited.

The biggest challenge in this section was Rumichaca Canyon, which was supposed to contain a series of Class V rapids. After about 12 miles of flatwater, I portaged around the strainers and rapids within the two-mile long gorge, ending with the broken *rumi* (bridge) *chaka* (stone) landmark for which it is named. Below the ancient bridge the canyon walls opened and the river flowed fast and flat to the destroyed Ocac bridge, southeast of Ondores, where I met the boys and we camped at an ancient dairy farm. Being alone in my kayak kept me far more vigilant than I would have been otherwise, knowing there was no chance of rescue, should I need it.

Upamayo Dam. Photo by Jeff Wueste

The next day, I paddled through a few easy rapids before abruptly spilling out into the Malpaso Reservoir. After crossing the expanse, I portaged over the huge dam via a set of stairs on the upstream and downstream sides of the dam, then met the team at the guard station two miles downstream. From there, an exhilarating 10-mile run brought me to our camp on an abandoned golf course outside the town of Paccha, which is Quechua for "Earth."

Power station below Malpaso Dam. Photo by Erich Schlegel

After the expedition's slow start, I finally felt that I was on a roll. On this, my best mileage day, I shot through 87 miles of read-and-run rapids without a stopping, between Paccha and a community park just short of Jauja, where Erich arranged lodging on the floor of a closed-down café. This section of river ran along the paved road and busy streets and villages. The Mantaro ran polluted through the trash strewn banks of downtown La Oroya. Along one side were the gated communities of the mining engineers and managers, while the workers lived in shanties along the other side.

The Amazon from Source to Sea 133

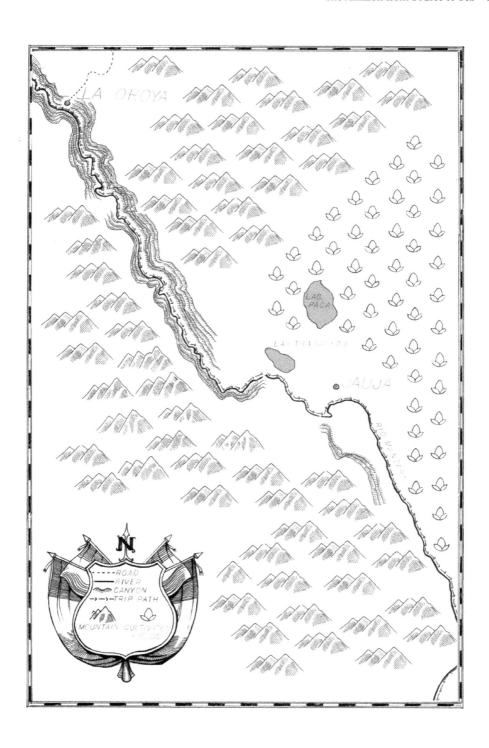

Ernesto "Che" Guevara and Alberto Granado left their homes in Buenos Aires, Argentina in January 1952 on the back of Alberto's 1939 Norton motorcycle. Their plan was to ride the wheezing single-cylinder machine all the way to North America, but after several wrecks and breakdowns they abandoned the ironically named *La Poderosa II* (The Powerful II) and stuck out their thumbs. They hitchhiked through Bolivia, Peru and Colombia in a nine-month adventure recounted in Che's memoir *The Motorcycle Diaries*.

Che was a 23-year-old medical student and Alberto, 29, was a biochemist. Though their families were well off, they opted for a vagabond approach, living off the kindness of strangers and working in leprosy clinics in exchange for food and shelter. Telling half-truths about being experts in leprosy, they eventually became quite proficient in treating the disease, though the time they spent with these abandoned patients and poor indentured workers was probably their most valuable gift.

When they reached the Peruvian highlands Che and Alberto were flat broke and hungry. Within earshot of a likely looking mark they launched into a loud and dramatic discussion of their woeful travels from Argentina. A kindly driver took the bait, offering coffee, food and a ride to Cuzco. They were in the back of his truck when they rolled through La Oroya without stopping. Che wrote, *"La Oroya is at an altitude of 4,000 meters, and from its unrefined appearance you can picture the hardship in a miner's life. Its tall chimneys throw up black smoke, impregnating everything with soot, and the miner's faces as they traveled the streets were also imbued with that ancient melancholy smoke, unifying everything with its grayish monotone, a perfect coupling with the gray mountain days."*

I can't mention La Oroya without making note of the significant feat of railroad engineering that made the place famous. In 1851, Polish expatriate Ernest Malinowski proposed an extension of the railway that had just reached Callao to the Jauja Valley. It was a scandalous proposition. Building on such a steep grade to so high an altitude had never been attempted in railway history.

After nearly two decades of study, Malinowski and his Peruvian colleagues settled on a modified route proposed by the North American engineer Henry Meiggs. Construction began in January 1870 under the direction of Malinowski and fellow Pole Edward Jan Habich. The railway opened in 1878 to the town of Chicla, then extended to La Oroya by 1893 and finally Huancayo in 1908. At the time the railway was the highest in the world, reaching a height of 15,692 feet. It held that distinction until 1984, when a rail line over a 16,640-foot Tibetan pass was completed.

Drag and Drop

"Never throughout history has a man who lived a life of ease left a name worth remembering."

Theodore Roosevelt

The Mantaro valley spread out downstream of Jauja and the river ran smooth, swift and easy all the way to Vieques, where I met Jeff and John at the exact point above Izcuchaca Canyon where we had launched with the Tigers three weeks before. Now the upper Mantaro was complete, so we stayed a night in the luxury of Huancayo before moving on.

Trashed banks of the Mantaro River below Huancayo. Photo by Erich Schlegel

The next day we leapfrogged ahead to run the 57 miles from the Tablachaca Dam to the Río Huarpa confluence. We were now well into the Andean dry season, and most of the Mantaro's flow was being diverted far downstream via the water cannon that had nearly pummeled Juanito and me to oblivion. Rocky wasn't the only one who was sure the riverbed would be dry. Piotr also made a point of insisting I carry my kayak the entire distance to preserve my claim to a full source-to-sea descent. Given that his expedition hiked the first 60 miles from the Apurímac source, I figured people must have given him some grief for not doing the same. Without the benefit of a visit to Peru, National Geographic geographer Juan Jose Valdes would later declare that this "de-watered" section invalidated the Mantaro River as the most distant source of the Amazon, simultaneously changing his definition of a river's source from what he used in an attempt to invalidate the source of the Nile River a few years earlier. Arriving at the dam, I was pleased to see that reality had defied all of their preconceptions. There wasn't a lot of water flowing out of the dam, but there was plenty to float my kayak.

The river below Tablachaca Dam was a series of shallow sections separated by occasional gravel bars, where I'd get out of my boat and drag it 20 yards through ankle deep water to the next pool. The route wasn't difficult, though my legs weren't used to the work of pulling the kayak. I became quite adept at finding water trails deep enough to float through.

A day of paddling and portaging brought me halfway to the village of Anco, past the remains of the 1974 landslide that obliterated the town of Mayunmarca and caused engineers to move the site of the Cerro del Águila Dam to our Dynamite Canyon, where flying boulders nearly crushed our team.

Che and Alberto spent time in Anco while hitch hiking north from Ayacucho through the center of the Andes. They were stopped by a landslide blocking the road and got underway again at nightfall, riding in the back of an open truck. *"To cap off our misery, night was closing in and a terrible rainstorm transformed the road into a dangerous river of mud. There was only room for one truck at a time, so those on the far side of the landslide came through first, followed by those on our side. We were among the first in a long line, but the differential on the very first truck broke when pushed too violently by the tractor assisting in the hard crossings, and we were all blocked again. Finally, a jeep with a pulley on the front came down the road, allowing the rest of us to continue on our way."*

Below Tablachaca Dam in the dry season. Photo by Erich Schlegel

That evening Erich joined Jeff, John and me in Anco, where we'd commandeered a seedy motel to retreat to after long days on the river. Erich brought good news; National Geographic wanted to make our expedition a cover story. The magazine planned to send Joe Kane, author of *Running the Amazon,* to shadow a short portion of our expedition and write the article. National Geographic also sent more money to Erich to continue covering the expedition.

Below the boulder fields of the Mayunmarca slide, the river opened up and flowed well, allowing me to double my mileage from the previous day. Expecting more low water and short portages I left my helmet and life jacket behind, which was a big mistake. Instead of the clear open channels I found immediately below the slide, the river narrowed, creating tricky whitewater through jumbles of bus-sized and car-sized boulders, which I had no choice but to run blind. More than once, I ended up in a space too skinny for my kayak and had to either back out or turn on edge to get through. Other times, I was surprised by six-foot drops hiding behind boulders. It was nerve rattling and I vowed never to paddle whitewater again without proper gear.

West negotiating the boulder fields. Photo by Erich Schlegel

The terrain took on the homey look of the Arizona desert, with saguaro cacti imparting a cowboy movie feel. The third day was easy flowing and non-technical, as I reached the Río Huarpa confluence just before noon. John was kind enough to drag my kayak up to the van parked a quarter mile upstream, near Puente Allcomachay, where we stocked up on sodas and junk food for the dusty ride back to Huancayo. Not long after passing through Mayocc, the van's transmission went out and we limped along the edge of the canyon in second gear. It was a long ride back to Huancayo, but at least I had time to think.

I'd paddled the 57 miles below Tablachaca Dam with only a few shallow-water drags the first day and none at all the next two days. The water level was well above what we had all predicted. Kayaking this section, along with all the accompanying evidence I'd gathered—including video footage, photographs, GPS tracking, live transponder footage and the accounts of expedition members and records kept by Tablachaca dam engineers—should have put the issue to rest. But facts are easily ignored when ambition and expedition politics come into play.

Rocky, ever the rational scientist, couldn't deny the data. Years later he conceded that water was flowing in the dry season through Tablachaca Canyon. More importantly, in his article revealing the new source in the journal of the Royal Geographical Society, he proposed for the first time in history a universal set of definitions for sources of rivers that does not rely on subjective opinion or constantly changing criteria.

West portaging over the shallows. Photo by Erich Schlegel

Rocky offered definitions for two distinct types of river sources, including the "most distant source," which is the point in a river system from which a drop of water will travel the longest distance to the sea, and the "principal" or "mainstem source," which is found by following the highest-volume tributary upstream at each confluence. In the case of the Amazon, the most distant source is the Río Mantaro and the mainstem source is the Río Marañón. Ultimately, and oddly enough, the constant presence of water in a river or stream became unnecessary in defining a river's course, given the intermittent and variable flow of all river and streams, with a definition that relied solely on the one constant trait, shared by all waterways: the streambed.

Rocky included in his paper a consolation prize for the Río Apurímac, calling it the Amazon's "most distant source of uninterrupted flow." He added this third definition as a nod to the historic significance of the Apurimac, while ignoring any historic significance of the Urubamba and Napo. For years Piotr has promoted his position that the Apurímac River, where his 1985 expedition began, remains the most distant source. This bit of mental gymnastics requires the invention of a new category just for the Amazon and no other river system on the planet. The idea being that the Mantaro is only temporarily longer during the rainy season, and that the Apurímac, though unequivocally shorter, retains the permanent title of "most distant." If you accept that argument, Piotr Chmielinski is still the first person to paddle the Amazon River from source to sea, no asterisk required; however, if you believe that all rivers should be measured using the same set of unchanging criteria, then the Mantaro is the most distant source of the Amazon.

Chapter Thirteen
Getting There is Half the Fun

"The struggles we endure today will be the 'good old days' we laugh about tomorrow."

Aaron Lauritsen

We had two days of downtime in Huancayo before Juanito and the new Tigers were to arrive, and spent the time resting, reorganizing gear (we were constantly reorganizing gear) and dumping food into my bottomless pit of a stomach. While I was growing leaner from long days on the river, Jeff and John complained of noticeable paunches from too many hours at leisure. We shoveled down burgers and fast food at the modern shopping mall, then found a wonderful café that specialized in fried chicken and one of the best chocolate shakes on the planet. The hostess propped us in front of a wide-screen television playing American music videos and kept our tab open until we'd eaten several chickens and drank chocolate shakes by the gallon. We then waddled back to the hotel, topping off at bakeries along the way.

While Jeff, John and I were all but invisible to the fairer sex, Erich's huge smile, square jaw and suave command of the Spanish language had them scampering out of the hills to make his acquaintance. More often than not, the three of us would simply fade into the woodwork, unnoticed by his suitresses *du jour*, amidst the cacophony of giggles and smooth flirtations.

During evenings at La Cabana, we spent plenty of time shoveling down pizza and hot rum tea while questioning Lucho about the Red Zone, a particularly lawless part of Peru that we'd soon be paddling through. Lucho listened thoughtfully, looking like Robert De Niro in one of his more rugged roles. He absorbed the conversation quietly and parsed out advice carefully. The news of Davey's shooting amplified our anxiety and we were seriously contemplating taking firearms. Lucho recommended we obtain a local guide who knew the lay of the land and could talk our way through any problems. Erich's friend Jonathan Shanin, an old hand with the conservation group Project Amazonas, recommended Iquitos resident César Peña, who knew the upper Amazon well.

Juanito and the new Tigers arrived late that night. Studying maps over coffee the next morning, Juanito announced that the first 30 kilometers of the lower canyons would be the hardest whitewater we've encountered thus far, which was a mighty bold statement. Though I knew nothing about the new Tigers, Santiago, Daniel and Chava, I was more confident in my whitewater skills after two weeks of kayaking on my own without serious mishap. The raft would have a different dynamic with Juanito's new oar rig and Erich on board with his cameras.

We got a late start, but Jeff was determined to get us to the launch point at Puente Chyquia before dark. He drove with wild abandon, slamming the hapless minivan into road mounds and craters, careening around single-wide hairpins beckoning us to an early death. The rest of us closed our eyes and tried to nap. I swear that overstuffed minivan stuck a triple gainer with a full twist and a half pike after hitting one particularly large bump.

Do not take a minivan into the Andes. Photo by Erich Schlegel

About halfway down the canyon we had two flats in quick succession. While changing the first tire, Daniel noticed the front brakes smoking. They were locked in place but we just kept going, trailing smoke and stinking up the van with burning asbestos. Just before the final 3,000-foot crazy switchback plunge to the put-in at Puente Chyquia, we stopped at the tiny village of San Antonio to have the tires fixed. The mechanic patched the holes with no electric or air tools while we ate eggs in the attached café. Confident in our spare and other air-worthy tires, we crept down the steep switchbacks, Jeff using the handbrake to save whatever he could of the still-smoking front brakes. A quarter of the way down our newly fixed tire went flat, so we installed the spare—our third tire change of the afternoon. Soon the spare started losing air.

As the sunlight waned an oddly civil debate ensued as to what to do. Jeff's idea was for everyone to go back to San Antonio and get both flats fixed. Juanito wanted to limp down to the bridge so the whitewater team could launch as soon as possible, be it today or tomorrow morning. John's idea was to dump the raft, kayaks, crew and gear right there on the road, several miles from the river in the middle of the wilderness, turn the van around and get the tires fixed in San Antonio.

Queue the awkward silence.

John was quite serious. At first, we opted to turn around and all head back up the mountain to get the tire repaired, then Juanito urged us to reconsider, as Jeff and John had plenty of time to get the tires fixed but the whitewater team (a.k.a. "the expedition") was under time constraints. We opted to limp toward the bridge, as we were closer to it than we were to San Antonio. John wasn't happy with this plan and expressed how this would be huge hassle for him and Jeff, without mentioning the remaining six of us (a.k.a "the expedition") who had several days of uncharted whitewater to traverse.

As we crept down the mountain, Juanito suggested that we all walk, to lighten the load on the deflating tire. John stayed in the van while Jeff drove and the rest of us enjoyed a truly pleasant evening stroll, given the circumstances. Eventually, the walkers caught rides to the bridge on passing trucks and Erich hitched a ride back to San Antonio with one of the flat tires, then caught a ride back with the repaired tire, arriving well after dark.

In the meantime, Barbara was on the horn with the rental car agency about the brake situation and all the flat tires. A company representative told her the van wasn't supposed to be driven on dirt roads. She pointedly asked if he'd ever in his life driven anywhere in Peru outside of Lima and he prudently backed down.

Given the late hour we arrived at the Chyquia bridge, we all bedded down with plans to launch at daybreak. John and Jeff fell asleep in the van, and well before daybreak, Jeff rustled me awake and quietly asked if the van was completely unloaded. He and John planned to leave immediately to get a jump on the day. I mumbled something vaguely affirmative, figuring Jeff would check the van for gear before taking off. I heard the van fire up and pull slowly across the bridge and went back to sleep.

Juanito woke me just as the eastern horizon was starting to glow. He reported that the van was gone and his gear was in it. Shit. We all jumped up and took a quick inventory to find several essential pieces missing, including paddles. I spent the next hour on the sat phone with Barbara and Jeff, arranging to get the gear back down the mountain as soon as possible. Jeff was already in San Antonio and made it clear that his priority was to get the tires fixed, regardless of the stalled state of the expedition. I tried to persuade him to pay someone to drive him down to the bridge with the missing gear, but he was adamant about staying with the van in San Antonio. He was tired and put out with John, who was increasingly disappointed with how the expedition was going for him. The missing gear was holding up all forward progress we could be making on the river. From my perspective, he had several days to fix the van while we were on the river, before we were to meet him in Ayacucho, so this delay was completely unnecessary.

I told him I'd made a mistake telling him we had all the gear out of the van and now I needed his help. I asked him point-blank if he thought his waiting for the tires to be fixed and taking no action to get the gear to us was the best thing for the expedition. Without hesitation or explanation, he simply answered "yes." This was the only time on the expedition when I felt Jeff had let the team down. I told him to wait in San Antonio, no matter how long it would take—it could be days—for the paddling team to figure out a way to get up the mountain, without a vehicle, and get the gear. Juanito was livid.

Eventually, Chava hitched a ride on a passing truck up the mountain. He grabbed the gear and some lunch for us all, and then hitched another ride back down the mountain. This took six hours, where Jeff could've had the gear to us in an hour. I was quietly furious, while Juanito stomped around, cussing and yelling obscenities.

The Lower Canyons

"If you are not willing to risk the unusual, you will have to settle for the ordinary."

Jim Rohn

We finally launched at 2 p.m. and quickly descended into a tight canyon. The Andes are actually two parallel mountain ranges with a high plateau (altiplano) between them. Below our put-in at the Chyquia Bridge, the Mantaro cuts a deep chainsaw gash into the eastern range, through which water flows with enthusiasm usually reserved for a kid fueled by moon pies and Big Red cola. The entrance to the lower canyons may as well be the Gates of Mordor, rising straight out of the valley and towering over the river. From a distance, against the briefly level river bottom, the slender entrance to the canyon appeared almost too narrow to pass through, except in single file. As the river shot through the gap it grew in velocity and for six miles we ran nearly continuous rapids with no chance to scout. The canyon walls were too sheer to climb and so closely spaced they blotted out the afternoon sun. We paddled in deep shadows, with the new Tigers leapfrogging around the raft with a wary eye out for danger.

With almost 400 miles of the Mantaro behind me, I felt more comfortable than ever on the whitewater. Erich, who Juanito had benched for most of the prior action, paddled furiously at each rapid until calmed by Juanito, who laughed more than I'd seen him laugh the entire expedition. He was back in his world. Before darkness completely filled the canyon, we found Rocky's old camp spot and pulled over with plenty of time for the Tigers to perform kayak ballet in the standing waves.

Chava, Santiago and Daniel had a different approach than the first set of Tigers. This new group was much more geared toward running safety for commercial raft trips, operating as a unit in close formation and with constant communication. When approaching a rapid the squadron would peel off into eddies one at a time, with a final Tiger perched downstream. As the raft passed, each Tiger peeled out, trailing the raft until everyone rejoined below the rapid. The first Tigers were expert gunslingers, operating independently and showing the route for the raft. The second Tigers were a well-tuned jet fighter squadron.

Juanito and West taking a dive. Photo by Erich Schlegel

Back in the States, Piotr pressed David about our schedule, in hopes of getting Joe Kane to interview us on the Ucayali River. Prior to this, Juanito hadn't really grasped the importance of the expedition, but he suddenly perked up and started helping Erich get the photos he needed, coincidentally including himself in every shot. Oftentimes, he made suggestions for shots we should get, much to Erich's entertainment. This was the best time I had with Juanito, who was hilariously mugging for the camera and choreographing Erich's work. "Fun" had entered the otherwise serious expedition vernacular for the first time.

Far from the anticipated deathtrap rapids, we ran several Class IV and two Class V rapids with relative ease. The last stretch of the Mantaro fit on a single topo map and the next phase of the expedition was soon to begin. I was in much better shape than when I began the expedition. My paunch was diminishing and my shoulder, though painful, was holding up. Under the mild influence of whiskey next to the campfire, far from the tension inherent with urbanity, things were looking good. We still weren't out of the gash that severed the eastern Andes, but tensions were now manageable as we bedded down with high hopes and the song of a cascading waterfall singing us to sleep.

Over coffee, Juanito told Erich and me about the quiet Tiger, Chava Loayza. He was raised in relative poverty on the banks of the Apurímac River, and lived in a house made of reeds until he earned enough money as a river guide to buy his family a piece of property in Cusco. The 22-year-old was modest and spoke little English. He was a fantastic kayaker and the smile rarely left his face. He was pure joy to be around.

Juanito and Chava. Photo by Erich Schlegel

Daniel Rondón hailed from Arequipa. He was the Ringo Starr of the group, friendly and sweet to a fault. His father owned a raft guiding company and his family ran the best pizza parlor in Arequipa; which says a lot, given that Peru, oddly enough, is quite renowned for its pizza parlors. In fact, a famous alleyway in Lima boasts ONLY pizza parlors. Thankfully, for my poor verbal abilities, Daniel's English was flawless. His skills and quiet approach to reading a river equaled that of Juanito, without the brooding seriousness. He hosted me when I visited Arequipa a few years later.

Daniel and Santiago. Photo by Erich Schlegel

Santiago Ibañez, also from Arequipa, was the joker of the group, exhibiting a dry and often goofy sense of humor. He got particularly tickled with my attempts at Spanish and coached me not only in the language, but in whitewater kayaking skills over the next three days. Two years later, a friend of mine from Texas met Santiago on the slopes of Mount Everest as they were in retreat from the devastating earthquake that killed thousands in Nepal. During the long hike from the mountain, the two just happened to strike up a conversation about their respective homelands and my name came up as a common link. Santiago later dated Juanito's little sister.

On our second day in the lower Mantaro, we ran through several tight canyons lined with beautiful rock striations. Frequent waterfalls plummeted hundreds of feet to the river, supplying us with fresh drinking water. For the first time, I started seeing bright colors, other than the drab greens, grays and browns of the highlands. Smells continued to elude me, though the nagging runny nose and intermittent dry hack persisted from our time in the higher a week before. The surrounding landscape was well watered and thick with vegetation, though we saw no villages or signs of people. As the river descended it made a lazy bend to the north, and as it turned back south towards the jungle the weather became warmer and the terrain more verdant.

West's helmet with satellite transponder, camera and stylish adornment. Photo by Erich Schlegel

That afternoon found us cruising quickly through wider expanses and smaller rapids. We didn't scout any rapids and once clear of the canyons, Juanito allowed me back in a kayak and had Chava take over the raft so that Juanito could train him in the art of rowing. It was good to be back in the kayak which was much easier on my damaged shoulder than the raft. The switch also gave me time with Santiago and Daniel, who were both extremely entertaining. Santiago taught me some Spanish words and Daniel laughed at my attempts. The group developed a light heart as the weight of the mysterious Mantaro lifted off my shoulders. By day's end, we had covered 37.6 miles, which was a pretty healthy run for a raft.

West deep in a hole (feather on helmet), Santiago about to roll back up, Daniel keeping a close watch and Juanito guiding the raft. Photo by Erich Schlegel

Almost suddenly, the smell of lush, thick greenery filled our senses. It had been a long time since we'd seen plants that weren't starving for water or threatening to stab us should we draw too near. Birds chirped and sang in large flocks nestled in the lush side canyons that met the river every so often. I never really noticed the loss of any natural odors, until they percolated across the river once again. For weeks, we existed in the cold, hard Andean highlands, where plants fought too hard to simply exist much less expend the energy to eke out a fragrance. Now, with water, sun and warmer air aplenty, the flora unleashed an olfactory kaleidoscope.

We found camp on a gravel beach on river right, well downriver from a suspicious-looking mining operation partway up an emerald cliff on river left. The Tigers figured this to be an outlaw camp for no reason other than what their guts told them. A cable with a raft ferry was stretched across the river to a well-built, but abandoned, hacienda on river right. The mine works, if that's what they were, trailed up the thickly forested mountain. There wasn't much activity, save for a man moving around here and there on a footpath. We paddled in anxious silence for another mile before pitching camp.

Juanito with his best Sendero Luminoso look. Photo by Erich Schlegel

Once below the encampment, Santiago pulled his T-shirt over his head and across his nose, imitating the hoods worn by the *Sendero Luminoso* or Shining Path guerrillas. He held his paddle like a rifle across his chest and glowered at us. After the laugh, I realized that he looked way too authentic and urged caution. The *Sendero Luminoso* had recently had somewhat of a resurgence, after being beaten back by the Peruvian military for the last two decades. The most recent incarnation of the Communist movement forged lucrative partnerships with drug traffickers, which the purely political rebels of previous generations once shunned.

Our camp, on river right, was on a long beach littered with…well, litter. A substantial evergreen forest spanned about 200 yards between the river and steep green mountains, and paths ran parallel to the river, well-used and marked with donkey droppings. I dreaded the thought of any visitors, especially at night. Erich and I strung our hammocks in the trees while the Tigers all slept on the ground near the river. The fire dried our clothes and warded off the evening chill, though we were all a bit more alert than usual. We dropped off to the sound of buzzing insects, chirping birds and squeaking bats. This place was alive.

We were now at around 2,000 feet elevation, having descended about 12,000 feet since the source. As the crow flies, the headwaters of the Mantaro were only about 140 miles away, but in terms of people, plants, experiences and water flow, the last 430 river miles had delivered us to another universe.

West, Santiago and Daniel kayaking through the cloud forest. Photo by Erich Schlegel

I neglected to bring my kayak spray skirt and paddle, so had to wrestle into tiny Chava's gear, which felt like a corset and the paddle a short toothpick in my paws. Still unable to roll, I swam twice, though never in a sizable rapid. Following lines run by Daniel or Santiago, I nailed the scary Class IV rapids, which pleased even Juanito, who was not one to dole out compliments. My injured shoulder prodded me to develop a masterful low brace, to which I owe my success in the larger rapids. Juanito never commented on my two swims, which led me to believe someone may have tipped him off about my injured shoulder, or maybe he just figured my poor whitewater skills were within tolerable limits, given the safety we had in place.

A pattern began to emerge in the Mantaro, where the water ran smooth and fast for two or three miles, then hit a series of short but significant rapids, all of which could be scouted or portaged, if need be. We ran them all fairly easily, after Juanito gave us the go-ahead and a game plan. Santiago and Daniel never ran me through anything I couldn't handle, but they would choose more challenging routes for themselves when opportunity arose. Once I noticed this, I often veered off on my own and took the easier routes, meeting the two below the rapid.

Santiago ribbed me about taking the *camino del pollo* (path of the chicken) of which I was quite aware and unashamed. Now and then I had to bust the hinges off the gates of hell to plow through a particularly deep hole in a rapid, determined to keep me far longer than I intended. For the most part, I had to remain healthy for another 3,800 miles to the Atlantic Ocean, long after Santiago would be safe at home eating some great Peruvian pizza and sipping pisco. Whitewater school would have to wait.

Erich wanted to get some close shots of the kayakers in the rapids. To do this, he requested we paddle right up to the stern of the raft as we barreled through the waves and holes. Once the roller coaster death ride began, Erich bounced around the raft with his monster-sized camera like a Cirque du Soleil performer, while I sweated and pissed my way through a watery hell, nearly getting sucked under the raft on several occasions when it stalled in a hole.

During one of the swift, flatwater sections that separated the rapids, Santiago sidled up to me. "Stay away from the raft," he said in a rare somber tone. "You can get stuck under there forever." Santiago spoke deliberately and looked me in the eye. From there on I backed off. Erich had enough photos.

West working through a rapid on the lower Mantaro. Photo by Erich Schlegel

Santiago and Daniel refill water bottles. Photo by Erich Schlege

On night three of the lower Mantaro we camped beside a massive two-story boulder, across the river from one of the ever-present jungle waterfalls gushing from the thick jungle slopes. We crossed some tipping point, where I knew the green would be with us from here to the ocean. From what I could tell, this was where the Amazon jungle began. This was a sure sign of progress, though perhaps I was the only one who felt it. Birds filled the sky, monkeys filled the branches, bugs filled the air and every available space of terrain was filled with something green, instead of the glorious, but barren, rocky existence of the mountains.

We stopped for an early camp, so that we wouldn't reach the confluence in the dark. Our team wasn't expecting us until the next day. This would make our final day on the Mantaro a mere 12 or 15 miles. I couldn't tell for sure because the topo maps had gotten wet in the so-called dry bag. That evening Chava Loaya and I shared the campfire. He wasn't a drinker, which left the chore to me, and we passed the time feeding the fire and eating snacks. Erich and the other two Tigers busied themselves setting up camera shots while Juanito, under the weather, went to sleep early. Chava, normally quiet beyond a smile or whisper, opened up and talked about his childhood on the banks of the Apurímac River. It was an upbringing full of love and support. He never went hungry and though he never had much beyond a full stomach, he was very proud of his family. I told him about what life was like in Texas, and he finally began asking questions, seeming to marvel at the small things. He was intelligent and more importantly showed great insight and perspective for such a young man. I was fortunate to spend time with him on that riverbank, in the quiet of the evening, amidst the tranquil sounds of the waking night, in such a beautiful place hundreds of miles from the inane.

During the night, the temperature dropped, necessitating that I crawl inside the sleeping bag inside my hammock. Adding to the chill was the presence of humidity for the first time on the expedition. My nose was ill-equipped for the intensity of life crammed into every sniff. Even the air smelled great. Every pore in my skin strained to absorb the olfactory blanket covering every molecule of space. If ever an orgasm of smell existed, this was it—all consuming, coming in waves, unstoppable and commanding complete attention. After the frozen mountains and arid desert, this mind-blowing orgy of scent was almost too much of a good thing. Perhaps a safe word would have been appropriate.

Amazon at night. Photo by Erich Schlegel

West, Daniel, Erich, Santiago, Chava and Juanito on the lower Mantaro. Photo by Erich Schlegel

The next day the river was void of all but the smallest rapids. I spotted my first long tailed parrots and flocks of colorful parakeets. We gawked at troops of spider monkeys flying along the treetops at amazing speeds, keeping a wary eye. *Oro Pendula* nests hung as their name suggests, high in the branches of the towering trees. These black birds had wide gold bands on their tails, which contributed to the "oro" part of their name. Green parrots flew in huge flocks, flitting and squawking from treetop to cliff top in wild unpredictable whims.

Steep mountains were covered head to toe in vines and greenery, blowing wonderfully in the occasional warm breezes that swept through the canyons and river valley. The Mantaro canyon spread out a bit, winding toward its confluence with the Apurímac. Tightly held walls relaxed into widening meadows and long beaches.

Just before mid-day we came upon the first village since departing the highlands. The houses, people and structures held quite the contrast from their mountain cousins, three days up river. River dwellers had an entirely different existence and their bodies exhibited such. The river held no place for the short, stout Quechua bodies adapted over millennia for thin air and harsh winters. Now, skinny children and adults played and fished in the river, most shirtless and all with huge smiles. Emotion was a rare thing displayed in the highlands.

Juanito steered the raft toward the bank, where he took on a small group of happy kids for a ride across the river. Though the river was swift, no one seemed overly cautious or concerned about being swept away. In the U.S., there would be warning signs, guards posted, serious steel railings erected and an army of lawyers ready to sue anyone and everyone if a kid happened into the water. Here, it was recognized simply as part of the world, neither good nor bad.

We all talked with the kids in Spanish, which gave them plenty of reason to laugh at me. They stared with wonder at me, Erich and Santiago, who were fair-skinned and fair-haired. He had to keep reminding them that he was, like them, Peruvian.

After passing our estimated 15-mile mark and not reaching the confluence, I reluctantly pulled up to some guys working on a large wooden boat to ask the distance. The river valley widened at this point, with more than a mile between the banks and the nearest mountains, which were no more than tall hills. The boatmen reported the confluence to be only 15 minutes or 4 hours or 3 days away. It was nice to have that all cleared up. One fellow offered to give us a ride to the confluence for a few *soles* and promised it would only take 5 minutes, warning that it would take us several days to get there in our boats.

The river widened even further as we came upon two men swimming across the river using empty plastic bottles for flotation. I paddled up to see if they needed help only to find they were stretching a net between them, measuring about 40 yards long. They pulled the net across the flow to catch fish moving with the fast current, then swam the tangled catch back to shore.

Further down, we came upon a couple of other guys doing the same, though they had made two rafts with two truck inner tubes and pieces of plywood. This, I figured, was the sort of contraption Davey DuPlessis, the vegan adventure tuber, was piloting when he was shot on the Ucayali a few weeks earlier.

At around noon on September 22, 2012, after a total of 39 days and 529 river miles, we finished the second successful descent of the Mantaro River and paddled straight up to César Peña, who was waiting for us on a gravel bar in the middle of the river. He arranged for a boat to pick us up in 30 minutes, which never arrived, of course. Two hours later he flagged down another boat, which took us upriver about 20 minutes, to a shithole masquerading as the village of Puerto Coco. From there, we loaded our gear onto a van, arranged by César, for the ride up to the mountain town of Ayacucho, where we would prepare for the final 3,800-mile push to the sea.

Waiting for our river taxi at the confluence of the Mantaro and Apurimac Rivers.
Photo by Erich Schlegel

Part Two
The Jungle

Chapter Fourteen
Out of the Frying Pan

"This ain't no upwardly mobile freeway
Oh no, this is the road, said this is the road
This is the road to Hell."
Christopher Anton Rea

Our driver proved his mettle shortly after departing Puerto Coco. While the 20-mile stretch of road to San Francisco, up the Apurimac river valley, was nominally paved, the bridges were narrow, allowing passage for only one vehicle at a time. When confronted, our driver aggressively nosed up to the vehicle, no matter what size, and honked like crazy, screaming and gesturing out his window for the imbecile to back the hell up. His name was Manuel, but Juanito addressed him as Papi ("Poppy"), Spanish for Dad. The nickname stuck.

If the frenetic honking and hollering didn't gain the desired result, Papi threw open his door and rushed the perpetrator, yelling and frothing at the mouth and floridly waving his arms. The screaming escalated until the opposing driver convinced the growing row of cars behind him to back up and allow this unhinged lunatic to pass before he started tearing trees from the ground and gnawing on tires. Once back in the van, Papi resumed his calm conversation with his co-pilot, Lance, without missing a beat.

We crossed the Apurímac at San Francisco, zipping by sandbagged machine gun nests manned by skinny young men in oversized helmets. The military presence reinforced my decision to gather the team in Ayacucho, far from whatever trouble these khaki-clad adolescents guarded against.

As we left town, and the last bit of pavement until Ayacucho 200 miles away, the road immediately began to climb. Unhindered by earthly constraints, Papi careened around hairpin curves as we quickly gained elevation. Even Juanito looked a bit nervous as we treated our anxiety with beer and rum. Within an hour we came to a construction barrier blocking the road. Papi launched into what critics rave was his best performance yet, standing inches from the kid guarding the barrier, showering him with spittle and invective. The young man, dwarfed beneath his hard hat and oversized fluorescent orange jumpsuit, never lost his look of stunned terror as Papi berated him non-stop for 10 minutes, to the awe of the crowd, which magically emerged from the surrounding jungle.

To his credit, the worker bee held his ground. The barrier wasn't going to move. Papi kicked the ground and spat up a storm before sulking back, defeated, to calmly report that we would have to wait.

"How long?" I stupidly asked, more out of habit than curiosity.

"Media hora," Papi responded, more out of habit than conviction.

Of course.

Everyone spilled out of the van amidst an impromptu swarm of capitalism. The team was quickly adopted by the women and children selling candies, breads, sodas and trinkets on the roadside. The Tigers were charming and Erich was the light of the show. He was propositioned a dozen times for matrimony or simply a honeymoon. I bought all their food and asked about locally made trinkets, but the only things on hand were factory made in China. Children clambered over the van, and improvised games of *fútbol* sprang up between the Tigers and the kids. The beer dwindled as hours ticked by and the line of vehicles grew. The shade provided by the overly lush and inappropriately steep mountain on which this farce of a road was perched cast the scene in a pleasant hue. Kids with anime eyes, having sold all their goods, looked up with palms out, begging for more money.

Andean jungle traffic jam. Photo by Erich Schlegel

West on the sat phone with Barbara explaining the situation. Photo by Erich Schlegel

Just before dark, the barricade was lifted. Papi fired up the battle barge and immediately bolted past the few cars he deemed unworthy of sharing the goat path. He bounced the van through massive dips and around steep berms created by the prehistoric earth-movers. Rarely did a mile go by without Papi slamming on the brakes and pulling in the mirrors to squeak past some ominous constriction caused by an errant landslide. Without speaking, we all instinctively slid to the side of the van farthest from the edge to lessen any undue pressure on the tires nearest the 2,000-foot plummet. I'm sure it was only our communally shifted weight that kept us from plunging to our deaths.

Juanito begged Papi for another latrine break and to restock our depleted supply of alcohol, which was needed in greater dosages to combat the terror. With ears popping and temperature dropping, we gained elevation in the darkness of the encroaching cloud forest, overtaking any vehicle traveling at a reasonable speed. Papi plowed full steam into the plumes of blinding dust in their wake, laying on the horn until slamming on the brakes inches from the offender's bumper. With each abrupt deceleration, the clutter of empty bottles on the floorboard rattled violently, breaking the stunned silence in which we all cowered. Defying physics and common sense, Papi screamed past, oftentimes on the precipitous death-side of the road. We pleaded in at least three languages for restraint, but all evidence indicated Papi suffered complete deafness.

Villages materialized abruptly out of the jungle darkness. At one small town, with a once-paved street and formerly solid colonial buildings, Papi tortured the brakes in front of a gated military checkpoint. I rummaged in the dark for the Department of Tourism dazzler as Papi fired up his jets. Before the uniformed teenager with an AK-47 could saunter up to the driver's door, Papi launched himself out of the van, cussing and screaming in his face.

The crowd mingling in the flood lights on both sides of the barrier froze, as did the soldier. Papi regarded checkpoints with the same disdain reserved for any vehicle on the planet not driven by him. The kid backpedaled as Papi shoved past him and rushed the open door of the adjacent building. In less than a minute, with great waving of arms and obvious irritation, Papi stormed from the building toward his open van door. Just before reaching the van, he tossed a couple of coins at the young soldier and demanded *"¡Arriba!"* with a flick of his hand. The pipe gate was immediately raised and we jolted back to top speed before Papi closed his door and quietly resumed the conversation with his co-pilot.

Papi at a single lane bridge stand-off explaining to another driver that he does, in fact, own the entire fucking road. Photo by Erich Schlegel.

The van bottomed out more and more frequently as the ruts grew deeper. Juanito pleaded in a meek voice for Papi to slow down, at least where the goat path thinned and the edge was steepest. His request was met with maniacal laughter and willfully ignored. In the dark, road workers in yellow coveralls stumbled home in small groups, kids played in the streets and various animals wandered about. Papi took great joy in aiming for pigs and children. Fortunately, all the dogs, pigs and kids had well-honed skills, dodging and jumping out of the way, centimeters from sure death or injury.

Then, there were the kittens. Who in their right mind would want to run over kittens? The van roared between a tight row of mountain shacks, hitting every pothole and slamming tsunamis of muddy water against slat walls and tin roofs. Through this murky chaos the dimmed headlights lit up a covey of cuteness in the form of five or six sweet bouncy balls of innocent fur playing in the road with a piece of something only a group of kittens would find exciting.

Papi stomped the gas pedal through the floorboard. Those of us staring in drunken horror over the bridge of the starship screamed sundry warnings and epithets in our respective languages.

"*Gatitos! Gatitos! Gatitos!*"

"*Cuidado! Cuidado! Cuidado!*"

"*...huh... ¿que pasa?*"

"Watch out!"

With no change in velocity, Papi hit the horn. At the last second, the kittens chased a bug off the road, near the wall of a house. We skidded by, missing the bundles of adorability by inches. Papi clearly didn't want to wait in line at the gates of hell.

Our hangovers grew stronger as we climbed into the mountains and the temperature plummeted. Papi treated his claustrophobia by keeping his window rolled down halfway, which shot a wind tunnel of freezing air into the jostling cavern in which we were trapped. Juanito's pleas resulted in Papi moving the window up one inch. My own pleas went unheeded. Finally, at one of the numerous urine stops and with the interior temperature of the van near freezing, I pulled Papi off to the side and stated in very slow, very clear, very broken Spanish that the goddamned *ventana* had better stay fucking *cerrado* until we get to the goddamned *ciudad* or Papi wasn't getting any fucking *soles* from me because we were *muy cansado* and *muy frio,* and I'm not going to put up with anymore goddamned freezing wind. *¿Comprendes?*

He got it.

Before descending into the Ayacucho valley, we pulled over for one more desperately needed latrine break. Papi and his co-pilot had bladders of iron, but then again, they weren't downing pints of beer. The lights of Ayacucho in the valley below sparkled against the charcoal-black mountains. The stars above were fistfuls of diamonds scattered on black velvet.

We sucked in bucket loads of fresh, cold air and dispensed gallons of spent beer down the dark mountainside. We stretched and sauntered back to the van, where Papi impatiently gunned the engine. Before we could slide the door shut, he lurched onto the newly paved road. In unison, we all began yelling at him to stop, as Santiago was left behind in mid-stream. Papi ignored our pleas, until Juanito was able to get the message through his deaf ears. He put the brake pedal through the floor, throwing us all forward and sending all the bottles slamming into the metal wall that separated the driver from us cattle. Santiago came running up, cussing and zipping his pants in mid-trot, then Papi jolted us back up to warp ten before Santiago could find his seat. The descent into the city was quick and littered with speed humps, which Papi, of course, treated as personal challenges.

Chapter Fifteen
Ayacucho

Rest and recovery in the death corner.

In 1952, Che Guevara and Alberto Granado rode in a cattle truck from Chincheros to Ayacucho, on their way north from Cusco. Battered by cold wind and drenching rain, they rode free, in exchange for watching after 10 young bulls with the young helper as the truck meandered through the rough mountain roads. Throughout their journey, Che took note of the learned helplessness worn by the indigenous to their plight. The cattle, weary of the effort to balance on the urine and feces, simply laid in the filth.

Tiny parade participants. Photo by Erich Schlegel

Che and Alberto spent only a few hours in Ayacucho, visiting two of the 33 churches in the small town, which is well-known for the battles Bolivar fought on the surrounding plains. Che passed through Ayacucho well before Abimael Guzmán founded the *Sendero Luminoso* movement here and before Che became a pivotal figure in the Cuban revolution. While both men later sought to shift power to the people through armed rebellion, Che leaned toward guerrilla tactics while Guzmán embraced the practices of Mao and the Khmer Rouge. Guzmán, inspired in part by Che's adherence to leaders who are of the people, was keen to expand on the concept and take control from the bottom up by winning hearts and minds. Che, less patient, wanted immediate revolution, from the top down.

Every school turns out for Sunday celebrations. Photo by Erich Schlegel

Working from his post at the university in Ayacucho in the 1960s, Guzmán was able to inject Maoist teaching into local schools. He later moved to the farmlands to establish a miniature economy, which was easy to do in this region far removed from the rest of Peru. His goal was an agrarian state in which the power was in the hands of the family farmer, a vision based on his reading of the traditions and myths of Incan culture. The movement became popular among the impoverished and working class, with many comfortable college students taking up the cause, which eventually became an armed rebellion. Like many movements with idealistic beginnings, Guzmán and his followers eventually expanded their violence to include the poor populace they originally sought to help.

Sunday is when the whole town joins in a parade. Photo by Erich Schlegel

During our brief stay in Ayacucho the sounds of revolution were but an echo, yet the *Sendero Luminoso* was still at work in the mountains between the city and the lowlands of the jungle through which we had just traveled. Near Ayacucho a year after our expedition, the government carpet-bombed a *Sendero Luminoso* hideout in the country. Ironically, or perhaps coincidentally given the excitement of our drive up from the lowlands, *"aya"* is Quechua for "death" and *"kuchu"* means "corner," reflecting the bloody battles that took place in the valley during Peru's war of independence.

Ayacucho plaza del armas on every Sunday. Photo by Erich Schlegel

Jokers to the left of me. Clowns to the right. Here I am... Photos by Erich Schlegel

Eventually, even the most ardent partiers have to quit. Photo by Erich Schlegel

Thanks for Flying Fuego Airlines

"If black boxes survive air crashes—why don't they make the whole plane out of that stuff?"

George Carlin

We spent the next three days in Ayacucho, resting and gathering teammates Pete Binion and Ian Rolls. Pete stepped up not only to fill Jimmy's vacancy down to Iquitos, but to also complete the route to the Atlantic. Such was his dedication that he said he would quit his job to participate in the expedition, if need be. Jeff and John drove from Huancayo to meet us with the three sea kayaks we would use on the flatwater; two tandems and one solo. John refused to drive, but helped in his own way by offering tips, analysis, criticism and complaints about Jeff's driving. The pair were supposed to paddle a tandem kayak together to the Atlantic Ocean, but Jeff had enough of their togetherness after a month in close quarters. He purchased one of the tandem sea kayaks from me at the onset of the expedition, which gave him the right to do with it as he pleased so he claimed Ian for his paddling partner. That left John and Pete in the other tandem kayak, and me in the single.

We also rendezvoused in Ayacucho with Jason Jones, who was able to join the support team after Texas authorities finally determined he hadn't actually tried to kill his girlfriend. Jason and I met in college 30 years earlier when he was on the Taekwondo team. Though I wasn't on the team the coach allowed me into the evening workouts. He was concerned that his students had grown accustomed to Taekwondo moves, leaving them vulnerable to fighting styles that weren't in the books. Jason and the team welcomed me into their fold and I got a group of friends with whom I could walk quite safely down dark alleys, for the mere price of getting the crap beat out of me on a weekly basis.

After college, Jason went into the Marines and found his way into jungle combat training, then bumped over to the Air Force doing something that involved giant transport jets. He came out with an unsettling comfort around jungles and firearms, an encyclopedic knowledge of big jet engines and a nagging case of PTSD. Rarely without a smile on his face, a Hawaiian print shirt and an adult beverage in his paw, Jason was the ultimate love child of Chuck Norris and Jimmy Buffet. There really was no downside to him joining the team, save for the pesky police investigation for attempted murder, which had simply been a misunderstanding.

A week before his flight to Peru, Jason's lovely long-term girlfriend Chris had taken ill suddenly. Emergency room doctors suspected arsenic poisoning and contacted the police, who called on Jason for a chat. After more tests, the good doctors finally figured out that Chris had a congenital anomaly that was easily diagnosed, which got Jason off the naughty list. He was allowed to leave the country just in time to join us in Ayacucho.

Rattled and worn out from five flights and three days of travel, Jason shuffled onto a crowded jet and settled into a window seat for the final two-hour hop over the Andes, from Lima to Ayacucho. Once the plane leveled off at altitude Jason glanced out the window, where a flap on top of the engine immediately caught his attention. It wasn't supposed to be open during flight. Most travelers wouldn't pay much mind, but most travelers don't have extensive experience with large jet engines. Jason knew exactly what the flap was, and why it was a very bad sign that it was open.

Jason pushed the overhead button directly wired to the sphincter-tightening muscles of every flight attendant around the world. He knew well the implications of pushing the button from his flight-attendant girlfriend—who, for the record (and it bears repeating), he had not tried to murder.

"Yes, sir, may I help you?" Flight attendants in Peru speak perfect English.

Jason kept his voice low so as not to alarm other passengers.

"That flap on the engine is supposed to be closed. Will you please alert the pilot?"

The professional bent down to look out the tiny window as Jason pointed out the flapping sheet metal.

"That's normal, sir. It's okay." But her eyes betrayed her. "May I get you a beverage?"

"Look, I was with the U.S. Air Force and I know a lot about jet engines. This is not normal." He kept his voice low, but his tone left no question that a very large quantity of feces was headed for that turbofan.

"I'll inform the pilot, sir. Would you like a beverage…on the house?"

"Yes, I'd like a Jack and Coke, please." Hell, it was morning in Lima but afternoon somewhere. Within seconds the drink appeared and the flight attendant made her way toward the cockpit.

She returned, even quieter than before, but with perhaps a bit more of a restrained expression, she whispered to Jason, "The pilot said that everything is okay. Can I get you another beverage?"

Jason weighed his options, which fell well shy of abundant.

"That engine is in serious trouble. There really aren't supposed to be pieces of metal sticking out from the engine. Please make sure the pilot knows this. I'll talk with him if you like."

She lowered her voice even further. "The pilot is aware of the situation, sir. Can I get you another beverage…on the house?"

Jason figured the pilot was as fond of his own life as Jason was of his and that the airwaves were abuzz with all kinds of chatter about the situation. He quickly looked around to measure up the first passengers he'd have to eat if they crashed in the Andes. Perhaps a liquid brunch was in order, indeed.

"Yes, please. Make it a double."

For the remainder of the flight, the attendant responded with amazing speed whenever Jason raised his cup. Their relationship had flourished to the point that he no longer had to qualify the request with a "double." Upon each delivery, the attendant made furtive glances out the window, though her fixed professional smile never wavered under her furrowed brow. The world went on as if the airliner wasn't about to be vaporized in an immense fireball at 22,000 feet.

Airplanes landing at Ayacucho must thread the needle between several jagged mountain peaks before settling onto a runway on a barely flat sliver of real estate tucked in the valley. As the plane slalomed toward touchdown, Jason downed his last (*last?*) Jack and Coke and strapped himself in for the show. He and his personal flight attendant exchanged knowing glances when she came by for his cup.

The airplane descended toward the runway, cocked sideways against the buffeting crosswind. When the plane was at cruising speed, air rushing past the engine kept it from bursting into flames, but as the aircraft slowed for landing, flames leapt from the errant engine flap. Jason unbuckled his seat belt and eyed the escape hatch.

In spite of the crosswind, and wing flambé, the pilot set the plane down like a pro. As soon as the rubber hit the tarmac a tornado of fire engulfed the engine completely. The pilot smoked the brakes, finally stopping at the very end of the very short runway where two small fire trucks were waiting. The intercom remained silent as fire crews raced to the burning jet engine. Within a minute the engine and flames were smothered with white foam and the passengers departed without a word from the crew.

Though a beautiful city full of archaeological treasures and splendid architecture, Ayacucho has remained off the well-trodden tourist trail due to its distance from the Disneylands of cultural archeology at Cusco and Machu Picchu. The town has developed quite a bit since 1903, when former British Army officer Ambrose Petrocokino trekked across the highlands from Cusco to Lima. He was not impressed with his accommodations.

> *"There is no service, and no water except a fountain in the street a block lower down. I may now mention that no houses in Ayacucho, or other higher Peruvian towns, are supplied inside or outside with "sanitary arrangements," so the early riser has to be wary when he walks the streets, else he will fare as did Henry IV, in Paris, when going to mass, at the hands of a student, but here at the hands of all the "ladies."*

The quaint cobblestone streets are now clean and quite void of any human effluent, and the population has grown from about 2,000 at the time of Petrocokino's visit to more than 150,000. The 33 magnificent post-colonial churches represent each year in the life of Jesus Christ, a significant figure in the conquering nation's religious belief. The nearby Pikimachay area contains archaeological sites dating back at least 15,000 years, along with extensive remains from the Huari, Chanka and Nazca cultures that preceded the Inca by thousands of years. While Machu Picchu is a wonderful thing to behold, it is a rather recent addition when compared to other human history that existed prior to the Incan and Spanish empires.

We were entertained by several parades throughout the town, remarkably similar to the scene Lt. Henry Lister Maw of the Royal Navy, described in 1827.

> *"Passing through the streets, we met several groups going about in masks, performing the old Indian dance, handed down from the time of the Incas; the music consisted of the ancient drum, and a kind of pipe, or flute. When they stopped to dance, it was in an irregular circle, and there was more noise than either grace or harmony in the whole performance. Knowing us to be strangers, they wanted money to buy chicha (a fermented liquor made from Indian corn,) and afterwards wished to pay a compliment by dancing round us. We declined the honour, and returned to the farmer's house."*

Of the many beautiful towns with rich histories and proud people in the Andes, I found Ayacucho to be one of my favorites, worthy of many repeat visits to enjoy the food, music, culture, art, architecture and climate.

Jeff, West, Erich and John at the plaza del armas in Ayacucho. Photo by a friendly waiter

Jason stumbled out of his taxi to meet us at our hotel, deservedly well-lit well before noon.

The Lucky Chicken

"Do ya' feel lucky, punk?"
Harold Francis Callahan

The drivers we hired to take us back down the highway through hell looked as if they'd just walked off a playground. We immediately dubbed them Justin and José Bieber as we loaded their van with far too much gear.

Before they left Texas, I reminded Pete and Ian to bring nothing more than a carry-on pack and their paddles, because we had all the camping gear, food, and tools they could ever want. Pete showed up with two huge duffel bags and a hard-sided paddle case the size of a set of golf clubs. Before boarding their flight in Texas, he had to hand over a pair of pliers to airport security. "That's okay," he told Ian. "I have two more pair of pliers in my checked bags."

Though we weeded out what we could from Pete's gear, he still had his tools, a full set of commercial pots and pans, silver and china place settings, a new washer and dryer, 12 bicycles and a set of radial tires. All of which would have to fit in his kayak with John's gear.

The team loading kayaks and gear in Ayacucho for the trip down to the Amazon.
Photo by Erich Schlelgel

With two drivers, five grown men, way too much gear crammed inside the van and three kayaks strapped to the roof, we headed out of Ayacucho late in the morning. On the edge of town, the Bieber brothers pulled over to allow José to quickly procure a live chicken from an adobe hut. He set the chicken between Justin and himself and we sped off.

I asked about the chicken, but they couldn't hear me over the *bop bop bop* of the cheesy pop music screaming to get out of the radio. I just wanted to be clear that I sure as hell wasn't paying for the chicken's ride. I had been bled enough.

The boys had the same "no defeat, no surrender" approach perfected by Papi when it came to oncoming traffic, though Justin scored much lower than Papi on the Knievel-o-Meter. Incessant rain and mist made the muddy precipices even more exciting than our uphill foray three nights earlier, though the cloud forest was quite spectacular in the light of day. Green drapes of foliage blanketed the otherworldly steep mountains, through which the road carved a meager existence. Now and again, we dove to the canyon bottom, crossing amazing picturesque creeks over faint bridges, only to wrestle up the muddy passage through the clouds along the wall of the opposing massif.

Drivers helping Erich's van through a washed-out section of road. Photo by Erich Schlegel

In 2012, Peru recorded 129 superfluously labeled "significant earthquakes" ranging from three to eight on the Moment Magnitude Scale (MMS), which has replaced the old Richter scale as a measure of tectonic indigestion. With unsettling thoughts of earthen flatulence running through my mind, I peered down the inappropriately angled precipices inches from our balding tires on one side and up at the crumbling wall on the other. Our Latin Justin Bieber casually swerved around bowling ball, breadbox and softball-sized boulders as they rained down, presumably from pissed-off mountain trolls.

Perhaps it never occurred to the road engineers that carving dirt away from the mountainsides would cause the earth above to succumb to gravity. Any seven-year-old with a sandbox full of Tonka toys could explain this phenomenon, and I'm betting the engineers didn't sleep downhill of their own handiwork.

We were able to weave around most rockslides by leaning our weight away from the opaque lip of death. Inevitably, we came to a huge pile of recently displaced boulders that left no room for passage. We were the first in line in the climbing group, while the downward dog line on the other side of the slide was packed with buses and larger vehicles anxious to abide gravity's laws.

José Bieber stepped out of the van and absently set a rock behind the back wheel. He was never remiss in this, his one official duty. Immediately, the chicken went freaking nuts, crowing, flapping and squawking like crazy. As if the hysterical chicken wasn't enough, the skies opened and dumped a wall of water on us, further eroding the loose mountainside above and below our ledge. The unprotected edge was a crumbly mess of sand, mud and rock, dropping tons of debris here and there with little or no provocation. I was quickly beginning to understand the chicken's point of view. Justin and José scampered to and fro, consulting with other drivers as the lines of vehicles piled up, coming and going. Finally, a kid in sandals and shorts fired up a huge excavator and worked the shovel like a surgeon, pushing car-sized boulders over the edge and clearing a tight path for the traffic.

Though several boulders were still in our way, we rushed back into the van. The chicken glared down its beak at me. Justin fired up the rockets, shoved the van forward against the nearest boulder and laid on the horn. The huge shovel slid the boulder out of our way. The driver of the bus facing us, a mere amateur, was caught off guard. He ground his engine to life and inched forward as loitering passengers scrambled into the open door. With another boulder shoved aside, we inched forward as Justin pounded incessantly on the horn, insisting the excavator work faster. Squeezing between the final boulder and a 2,000-foot plunge to hell, we crept skyward. Justin pulled in his side mirror for that extra inch we needed between the passive-faced bus riders and the sheer drop. Though I had relieved myself over the edge mere moments before, the urgency to do so again swept over me, perhaps in an unconscious effort to make my body lighter.

Three hours later, we pulled into a *tienda* for snacks and to stretch. On cue, the chicken went bat-shit crazy. I pulled Justin aside and asked about the chicken. In somber and earnest tone, as if schooling a child, he told me, "This is a very dangerous road. You do not take this road without a lucky chicken."

Of course.

Upon depositing us, safe and sound, back in Puerto Coco, Justin made a feeble attempt at Amazon, shuffling us up 100 *soles* from the agreed upon fare of 650 *soles,* to no avail. Just to prove the shuffle was a matter of business-as-usual with no hard feelings, he drove back to us 10 minutes after departing to return some electronics I left in the van.

As for the fate of our lucky chicken, she was unceremoniously deposited at someone's house near the river, her luck having run out.

Fredy's Resort

"Don't you know there ain't no devil, it's just god when he's drunk."
                                                                Tom Waits

We were greeted in Puerto Coco by a guy named "Fredy," hugging us with moldy cigarette beer breath and a barrage of slurred Spanish. We couldn't shake him, in spite our repeated *"no hablo españols"* and *"no comprendos"* though we understood every word he garbled and spit. Diesel fuel, gasoline and the reek of burning plastic permeated the air.

We quickly set up our tents near the river behind Fredy's shack, which consisted of a log pole frame wrapped in black plastic sheeting and filled with drums of diesel fuel and gasoline. Fredy was proud of this shithole and lord of his domain. He tried to get us to stay up in his *hospedaje*, but we insisted on sleeping next to our kayaks. Fredy stumbled between each of us in turn, offering swigs from a bottle slick with drool.

"No, gracias." We did our best to shove him off on each other, claiming that the other guy spoke fluent Spanish and loved spit-tinged beer.

At one point, Fredy—in mid-sentence on a topic of utmost importance—handed his bottle to Ian, undid his pants, squatted and shat, pulled up his pants, took the beer from Ian and landed another swig without missing a beat. We had a winner.

We traded off guarding the tents and gear in the dark during dinner at the makeshift café next to the fuel shack. Fredy plopped down at our dirty picnic table, and continued his mumbling diatribe, replete with laughter at his own witty repertoire. The ubiquitous huge flat-screen television hung from the tin-roof by bailing wire, attracting huge flying bugs, which, like everyone else outside of North America, were fans of *fútbol*. Oftentimes, it was difficult to tell the *fútbol* players from the gargantuan insects making love to the screen. An unmuffled generator deadened all discussion, and the boom box vomited 1980s faux disco, adding to the ambiance. The chicken, rice and plantains were well done so we all partook, with the exception of Pete who announced a strong warning to us all against any local food. Those of us who had been in Peru for two months did our best to dissuade his suspicions, to little avail.

Departing Fredy's compound at Puerto Coco. Photo by Erich Schlegel

Fredy poured a thick cloudy liquid into dirty glasses, telling us it was *"agua de vida"*—water of life, sure to give us vim and vigor. The more he spit and slurred his incoherent excitement, the less we were inclined to indulge. Finally, to prove the swill was in fact quite safe, he reached across the table, drops of sweat flying off his flabby arms onto plates and silverware, and downed half my glass, then handed the remainder back to me with a fresh gob of drool creeping down the edge.

Check, please.

At that, we all developed a sudden fit of narcolepsy and found our ways back to our respective tents. At some point, deep in the night, Fredy stumbled over on top of Pete, who got the distinct impression Fredy was looking for a bunk to share.

Fredy eventually passed out, thunderously snoring on top of the fuming petroleum barrels inside his black plastic shack, next to my tent. I dug out my foam earplugs, which provided some relief. The damp morning came all too soon. Among a growing crowd of curious, way-too-comfortable onlookers, we made breakfast on piles of lumber and tried to shove too much gear into the three kayaks.

About this time, Erich and company pulled up in a van, having traveled all night through huge storms. Jason jumped out and proclaimed, "Y'all won't believe what we just went through!"

Chapter Sixteen
Shotguns on the Ene

Day 46, Mile 524

> *"Well, which is it, young feller? You want I should freeze or get down on the ground? Mean to say, if'n I freeze, I can't rightly drop. And if'n I drop, I'ma gonna be in motion. You see..."*
>
> Hayseed, *Raising Arizona*

Erich had yet to arrange for a pecky pecky to carry his crew and gear. I truly felt for him but at this point, I had no more fucks to give. If I remained in Puerto Coco, the amount of alcohol and violence would have been catastrophic. The circus was leaving town, I told Erich calmly. At least it sounded calm to me.

With a communal sigh of relief, we departed Puerto Coco in the rising fog, leaving behind the diesel and gas stench, heaps of burning trash and Fredy's hospitality. Finally, after weeks of whitewater, we were in our element. The five of us knew flatwater and long miles better than most anything else in our lives. Each of us, aside from Ian, had more than 20 years of ultra-distance racing experience.

Though overcast, the humid day allowed just enough sunlight through the opacity of the clouds to glow like a bank of fluorescent bulbs in a low-ceilinged cubicle farm. Truncated mountains covered in cloud forest gently constricted the Ene into a very pleasant river with a healthy current. It carried us into the heart of the Red Zone, a lawless expanse of jungle at the epicenter of the global cocaine trade. Known as the "VRAE" (an acronym made up of the first letters of the local rivers) this area is legendary for killings and strife over the past four centuries, fueled by exploitation of indigenous people by profiteers, government and outsiders and their resistance to them.

During our lunch break, John refused the can of tuna I offered, stating "I'm not going to eat that." Turns out he didn't like tuna. Our hatches were filled with cans of tuna, since it was a staple of the expedition menu we'd all spent months planning. He made do with crackers and peanut butter.

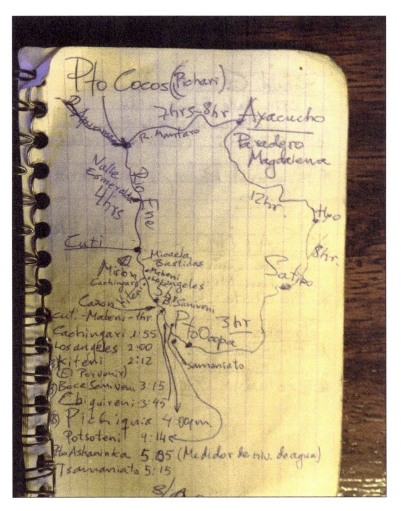

Lucho's map of the upper Amazon, including the Ene River. Photo by Jeff Wueste

Throughout the day, Jeff and Ian entertained themselves by sprinting in the occasional wave trains to see how fast they could run up the speed indicator on their GPS. John was in the stern of his boat, steering the tandem kayak with Pete in the bow. I had high hopes that John's negativity would soften somewhat in the presence of Pete, who was the elder statesman of the group and an old friend of John's. Pete was about as tough as they come, with a quiet, commanding demeanor I'd grown to respect when he and I teamed up to win the Texas Water Safari in a six-man canoe. He was a Marine in Vietnam, then spent years working the docks at the Houston ship channel, working in a municipal water department and running cattle on the prairie while raising a fine family and a handful of grandkids with his lovely wife, Becky.

Civilization dwindled down to nothing until we came to a tall bluff on river left, topped with neat white buildings, manicured lawns, radio towers and training fields. We exchanged waves with the young military recruits at the water's edge, who were washing their uniforms under the eye of their matron.

40 miles registered on the GPS when the sun dipped behind the mountains. The shade was welcome, but the mileage was discouraging. Every now and then we turned to look upriver to see if Erich's support team was in sight. This area was renowned for its superstition and fear of westerners, which had led to some recent killings of outsiders. Curious people stood and stared at us from a safe distance, their faces impassive despite our exaggerated waving and smiling.

"Hey there! Look at me! I'm harmless and friendly! Thanks for not killing us!" White people in the Red Zone never meant anything good for the locals.

With our water running low, we pulled over at the Río Catsingari where it spilled into the Ene. This small but fast-flowing river emerged from the greenery on river right, with a large stone beach extending several hundred yards upstream and down. Carefully landing upstream of the Catsingari river mouth, we set to work assembling water filters and collecting water into empty bottles. After a few minutes taking turns at the pumps, I glanced back to notice John and Pete unloading their camp gear. We hadn't discussed camping and this area was far too crowded. Kids and women across the river stared at us suspiciously as they washed clothes.

Pete walked up the beach and confirmed that he and John thought we were pitching camp there.

I went down to John to let him know we weren't camping yet, just getting water.

"First I've heard of it," he grumbled.

This statement became John's mantra throughout the expedition, beginning high in the Andes. John was in the habit of secluding himself in his hotel room or elsewhere away from the group when the rest of the team sought one another's company to form plans for the moment or day's ahead. He preferred Jeff's company, as they were longtime friends, but even Jeff was growing weary of John's disappointment in the expedition. We frequently hunted John down to include him, though this meant holding up the meeting or repeating everything we'd discussed. After a few weeks of this, we just didn't have the energy to accommodate his need for solitude. This, of course, led to him being inadvertently left out of discussions and group decision-making processes.

As the expedition leader, I should have talked with him early on about this concern and come to some mutual expectations, but with so many balls in the air, this particular duty ranked pretty low on my priority list.

"Well, that makes perfect sense, since this is the first we've talked about camping," I replied with curt calmness. A better man would have let it go, but his indictments of purposeful omission rubbed me the wrong way. Walking away, I didn't feel good about the exchange, but more notably, I didn't care.

While we packed to leave, some guys with shotguns trotted down the embankment across the river, heading toward a pecky pecky. Long before we entered the Red Zone we'd planned for situations like this. If approached, we would spread out and leave one boat to deal with the gunmen, allowing the other two kayaks a chance to get away.

I paddled away from shore, spun into the current and headed straight for the pecky pecky as it motored upriver. When I was about 100 yards from the rest of the group, about to exchange friendly greetings with the gunmen, they yelled and motioned for me to go back to shore. Feigning confusion, I asked "*Porque? Porque?*"—Why? Why?—which didn't work at all. On the upside, they weren't shooting at me. Having barely launched, tandems were back on shore before I could fight my way back up the strong current to join them. The gunmen had three single-barrel 20-gauge shotguns of an ancient but seemingly functional vintage. This meant they only had three shots before reloading. In the worst-case scenario, we had a fighting chance. It helped that we were a foot or so taller and outweighed them by 50 to 80 pounds apiece. My apprehension turned to irritation.

The *jefe* of the group was an anxious little guy showing classic symptoms of inebriation. We could manage this. A drunk with a shotgun had "redneck Friday night" written all over it. They stood too close to us and lowered their barrels, often resting the business end of the gun barrels on top of their bare feet. If someone yelled "boo!" they may have blown a foot off. The *jefe* banged on our hatches, insisting we must have *drogas* as he stumbled between the kayaks. He spoke a mix of Spanish and something else. He wanted to search everything, starting with the large dry bag lashed to Pete's deck. The *jefe* knew for damn sure it was chock full of *drogas*. Pete hauled the large bag up to the shore and dumped out all of his clean clothes, through which the *jefe* rifled suspiciously. He then went to the hatches, which must have been a clever effort to conceal our volatile cargo. By now, the gunboys looked bored. Perhaps the bright white kayaks, the National Geographic flags flying from the small poles on the sterns and all the bright sponsor logos clued them in on our purpose. Maybe they just weren't as drunk as the *jefe*.

Drunk gunman and gunboys on the Ene. Photo by West Hansen.

Hatch by hatch the *jefe* found no *drogas* yet continued his frantic search. A crowd of women and kids gathered for the show. Pete gave me a couple of indecipherable looks, which may have been his way of asking whether we should grab the guns and beat the crap out of these guys, or perhaps order an Egg McMuffin. I wasn't certain, so I waved him off with the subtle wince of a minor-league pitcher.

The *jefe* looked a bit scared, so we all remained calm as I introduced each one of us by name. This wasn't my first staring contest with a gun barrel, though I've never gotten quite comfortable with it. I bucked up the *jefe*'s status for the crowd, showing an embarrassing amount of respect and asking his forgiveness for trespassing on his land. After he searched each boat, he motioned for us to go. The two tandems took off while I hung around to shake his hand and make small talk while I packed my kayak. I thanked him for his time, wished him good evening and paddled down river to meet up with the team. Now, we were awake and alert. Nighttime was minutes away and we had yet to find a camp spot in this dangerous area.

We set up our tents about two miles downriver, after asking permission from a fisherman plying his line about a hundred yards upriver. He motored up to us after dark and told us politely to leave, as we weren't on his land. He recommended we find a *hospedaje* in Quiteni, a few miles downriver. The idea of paddling at night in the Amazon jungle held no affinity for us, but we packed up and headed out.

Somewhere in the next couple of hours, we unknowingly passed Quiteni while trying to be invisible to the pecky peckys lurking in the darkness. Finally, we found a set of sand berms that might hide us from the river. After dragging the heavy kayaks around the mounds, we noticed two guys fishing on a gravel bar, staring right at us. We bought the fish they offered for a few *soles*. Protection money.

We made a pact not to tell any of our loved ones about being held at gunpoint. They were all very concerned about our safety, knowing we were in the Red Zone. Barbara told me on the sat phone that Erich and his crew were camped at the military base we passed several miles upriver. His pre-arranged pecky pecky never showed, so he'd had to find another.

The stars that lead our way to the camp spot disappeared against the blazing glow of the rising moon. In the moonlight we could easily make out three or four huts on stilts between us and the jungle. We went to all this trouble setting up a hidden camp and ended up camping in someone's front yard.

I spent the restless night rushing naked between my screen tent and the latrine as the remnants of Fredy's supper made for a speedy exit.

Peeled Faces Among the Unconquered

Day 46, Mile 592

"It was one of those humid days when the atmosphere gets confused. Sitting on the porch, you could feel it: the air wishing it was water."
                                                                    Jeffrey Eugenides

With the sun rising above the emerald green mountains, Pete roused us with an abrupt, "Let's get moving, boys!" above the lush jungle din. It was good to be on moving water in the daylight, after the anxiety of our night run. Unaccustomed to the moist atmosphere, we tried and failed to dry out our sleeping bags and tents before packing them away. Little did we realize how much the dampness would become simply a part of our daily existence for the duration of the journey.

This beautiful section on the lower Ene River was sprinkled every five miles or so with villages where our joy-filled salutations were met with stone-faced stares. This area is populated by Asháninka and Shipibo people, who over the past few centuries have fought off waves of missionaries and would-be conquerors. This resistance has continued to modern times, with several reports of outsiders being killed or maimed in the months before our expedition. One such incident, a few weeks prior to us entering the area, involved a western schoolteacher who had lived in the jungle among the local people for several years. One morning he awoke to find several tribesmen on the opposite bank, armed with bows and arrows. He called out to them in their native language but received no response. After a few minutes, he went back inside. When he came back outside again, they riddled him with arrows. He barely survived to tell the tale. He didn't know why they attacked him, since he'd always had very cordial relations with them.

During their Amazon Extreme rafting expedition in 1999, adventurers Colin Angus, Ben Kozel and Scott Borthwick were shot at on several occasions and held for several hours in a village until an English-speaking teacher convinced a tribal leader to release them. After that, Angus and company rowed their unwieldy whitewater raft 24 hours a day to get through the Red Zone.

On the Mantaro, I had been able to drift periodically and rest my damaged shoulder, but the flatwater meant swinging the Epic double-bladed wing paddle for 12 straight hours every day. Every stroke was accentuated with stabbing pain in my shoulder. Overdoses of ibuprofen didn't really work, so I took to keeping my left arm unusually low, while maintaining proper form on the right. Having stared at my back for several thousand miles over the previous 20 years, Jeff noticed the issue right off the bat so I had to fess up about the shoulder.

We paddled unhindered to the Canyon of the Seven Devils (*Pongo de Paquitzapango*) infamous for seven huge whirlpools, each said to be big enough to swallow a passenger ferry into the bowels of hell. From a wide stretch of river, we took a sharp right curve and headed southeast into the narrow canyon. The forest softened up the cliff sides, with rock outcroppings and greenery along the steep banks. The river narrowed and the current increased, but not to the point of creating rapids. Desolate, clean and beautiful, like the American West before it was "American" or "West," the two-mile canyon was full of calendar-perfect scenery. This section of the Ene was awash with waterfalls spilling from the forest, and clean-flowing mountain streams washing over cool stones to mix with the muddy main river. In spite of the legends, we found the canyon quite sublime and without peril, at this water level.

This was the beginning of the Gran Pajonal, a remote region along the river's eastern shore marked by the last truly large mountains in the Amazon basin. The verdant jungle is impenetrable, save for footpaths. The lone road is a rock and dirt nightmare stumbling along the western bank, twisting into knots all the way to Puerto Prado and then on to Atalaya.

Far from industrial odors, we submerged ourselves into a thick aroma of freshly mowed grass on a hot afternoon, multiplied by infinity. Moist, fertile bounty sprouted from every possible chunk of dirt. Sunlight bounced off each leaf in unending shades of green, yellow and dark reds. Now and then, a thick yellow-green banana grove burst into existence right in the middle of the wonderfully disorganized mosh pit of vegetation.

Before becoming a wilderness strip-tease artist on the Discovery Channel, Ed Stafford took a few years intermittently walking the banks of the Amazon, a journey he recounted in his best-selling book *Walking the Amazon*. He was met with suspicion in almost every village from the start of the Ene River and was frequently harassed and held at gunpoint. Stafford was eventually rescued, admonished and befriended by a woman from Italy, who had gone native in her self-appointed efforts to protect the Asháninka. Even with a local Asháninka guide, Stafford was doused with buckets of dirty water, an act of protest from the village women who yelled and hissed at him. Governments, commercial interests and religious authorities, who believe they knew exactly what is best for the locals, still carry the standards brought by Spanish conquerors 500 years ago.

Our friend from Huancayo, Lucho Hurtado, has his own horror story about this part of Peru. He and a partner came to this area with some fellows interested in selling their ranch land several years ago. After looking at the land and returning to town, Lucho and his friend were held at gunpoint by a small group of men. Following a brief discussion, two gunmen were given orders to take Lucho and his buddy "down to the river", which was akin to taking ol' Shep out to the farm.

Blessed with an amazing voice and a fountain of intellect and optimism, Lucho bargained with the gunmen as they strolled through the jungle on the way to the river. Lucho surmised that the two gunmen might earn substantially less money than they deserved for this distasteful chore, so he proposed a gift of hard currency. A mutually satisfactory amount was agreed upon and Lucho and his buddies were given a reprieve, with the promise to disappear so as to keep the gunmen out of Dutch with their boss. The men fired two shots into the air, then walked back to report a job well done.

After dark, Lucho and his friend climbed fences and scrambled through gardens back to town, avoiding paths and thoroughfares. Word traveled fast that the ghosts of the two visitors had appeared here and there, flitting through town. The fugitives from injustice made their way out of the jungle in the bottom of a produce truck. True to his word, Lucho sent the agreed upon sum to the two gunmen and never returned to the jungle through which we now traveled.

Around mid-day, Erich, Jason, César and a boatman arrived in the elusive pecky pecky to shadow us the remaining 60 miles to Puerto Prado. Now and then, the boat pulled up alongside for César to yell, "This village is bad!" or "This village is good!" though we had no plans of stopping at any of them. As we paddled by Pitziquia on river right (the Asháninka side), a bunch of kids played on a small log raft in the shallows. Typical kid stuff—pushing one another off of the raft, laughing and cutting up. We did the smile and wave routine and for once the kids pointed, smiled and waved back to us. Within seconds, a crowd of adults on the opposite bank screamed hysterically at the kids, *"PELA CARA! PELA CARA! PELA CARA!"*

Immediately, the kids leapt into the water, abandoning the raft and swimming like crazy for the shore then sprinting uphill to the houses, having seen the real-life bogeymen of colonialism, evangelism, racism and capitalism that their parents had dutifully burned into their impressionable minds.

While being a pariah stung, I couldn't fault them their response. A century ago, a popular theory among the Asháninka was that outsiders used the blood of local people to make the grease needed to run their modern machines. These days, whenever someone ends up missing—which is an easy thing to do in the jungle—the go-to response is that a *pela cara* snatched them for their organs to transplant.

People in the Upper Amazon had never seen human beings with light-colored skin until Spanish explorers pushed into the Peruvian rain forests in the 1500s. Because the Spaniards' skin was similar to the color of bleached skull—and perhaps because their arrival was so often accompanied by death—the local people called them *"pela cara,"* which roughly translates to "peeled face." While some historical reports claim that westerners peeled the faces of locals for a variety of horrible reasons, the people we met say the label comes from our resemblance to skulls. It is a derogatory term sometimes used in jest, like "gringo," but we found it uncomfortable in any context.

Jason, Erich and César meet us below the Canyon of the Seven Devils. Photo by Erich Schlegel

Jeff and Ian hit their speed tests from time to time, while John and Pete lagged behind, paddling well below their abilities. John was not paddling in sync with Pete's easy stroke, which is the first thing you learn in marathon paddling and after 20 years should be as natural as breathing. Poor timing slows a boat considerably, but telling a veteran marathon racer that he's out of sync is akin to correcting the spelling of a Pulitzer-winning author. I wasn't sure how to broach this awkward subject, or the disconcerting fact that John and Pete's boat often strayed far from the others. While this was confusing and somewhat dangerous, we were moving forward so I let molehills remain for the time being.

The swift water of the Ene carried us through a Hollywood-perfect jungle setting with miles of beautiful, serene shorelines through which spilled beautiful streams and cascades of pure water from steep, uncorrupted mountainsides, often topped with lush pastures of unadulterated St. Augustine grasses.

Chapter Seventeen
Puerto Prado
Day 47, Mile 641

"People in small towns, much more than in cities, share a destiny."
                                                    Richard Russo

When Puerto Prado came into view, Erich and his team zoomed ahead in the pecky pecky to begin the process of securing a permit to pass through the Red Zone. The town lies a few hundred yards up the swift-flowing Perené River, which joins the Ene to form the Río Tambo. Jeff, Ian and I beached on a gravel bar at the confluence to form a plan of approach to the town, as we would have to paddle upstream against the heavy flow. Pete and John stopped short about 100 yards up the beach. I walked back up river to investigate. John said they stopped because he didn't know what we were doing and Pete's back was sore. This didn't answer why they didn't drift the remaining thirty seconds to join the rest of us, but I let it go. I reminded them of our plan to stay overnight in town to get the permit, a topic we'd been discussing for two days.

"First I've heard of it."

Pete remained silent. At this point, I was stretching to give him the benefit of doubt and truly wondered if John had a serious hearing impairment, but suspected his issues ran deeper.

By the time we clawed our way up the half-mile of fast-moving water to the edge of town, Jason secured bunks at the Las Brisas Hospedaje and organized a bunch of guys to haul our kayaks down the dirt main street for a few *soles*.

Ene River catfish. Photo by Erich Schlegel

The ubiquitous yard fowl kept a watchful eye on the kayaks until dinner, when their mission shifted from guard duty to entrée. Thankfully, the usual crowd of gawkers didn't linger. Two-stroke generators pock-pocked the breezy air, while impotent power poles lined the street. Three-wheeled mototaxis plied the dirt streets, but without the same fervor as in other towns. Puerto Prado was comfortably lethargic. The pleasant Las Brisas proprietress, Noame Bisaeas Bijualo, showed us to our cavernously dark and sauna-like upstairs rooms, each furnished with a foam mattress, mosquito net, pillow and single sheet. I was glad to have a handful of earplugs. The tectonic snoring at Las Brisas that night was the stuff of legend, or so I was later accused.

Jeff, John (blue shirt), César, Ian and Pete hauling kayaks to Las Brisas Hostel. Photo by Erich Schlegel

Erich and César took the pecky pecky further up the Perené to Puerto Ocopa, the seat of government in these parts. There they met with the *alcalde* (mayor), who presented them with a tourist map and an ornate document that gave us permission to travel through the area without being shot.

"So, here is your tourist map of the beautiful and unique sights to be seen and enjoyed as you vacation in the lush, tropical vacation land of the Asháninka, and here is your permit that allows you to do so without getting your head blown off. Please visit the souvenir shop on your way out. Buh-bye."

In 1849, Father Juan Crisóstomo Cimini paddled up the Tambo River with 150 heavily armed men determined to spread God's love to the "savages," or else. They proved no match for the Campas, who despite being outnumbered and armed only with spears and arrows, sent the surviving evangelicals packing.

Two years later, Herndon proposed to follow the same route upstream and then up into the Andes in order to blaze a trade route. Though he had far fewer men than the evangelizers, he figured his peaceful mission to increase commercial traffic through the Campa homeland would assure him safe passage. He never had a chance to test that theory, however, as he couldn't persuade anyone to accompany him up the same route, even at double pay.

Hotel Las Brisas, Puerto Prado. Photo by Erich Schlegel

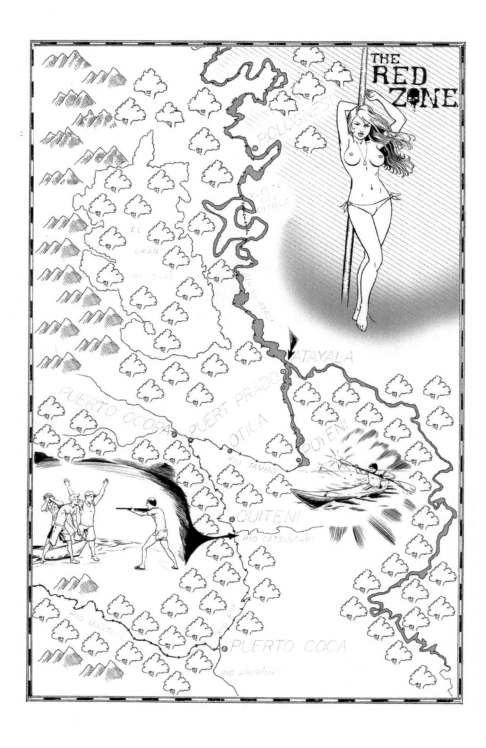

Badges? We Don't Need No Stinking Badges!

Poyeni - Day 48, Mile 692

"Look out Mama, there's a white boat coming up the river, with a big red beacon, and a flag, and a man on the rail."

Neil Percival Young

The mayor of Puerto Prado left us with strict orders to stop at checkpoints in Otica and Poyeni manned by *ronderos*, the local militiamen who are the closest thing to the law in these parts. The checkpoints cut the Tambo into thirds. The river is 108 miles long from Puerto Prado to Atalaya, where it joins the Río Urubamba and becomes the Ucayali. I naively figured we could cover the distance in one day.

Erich hired Noame's husband, Zocimo, to haul his crew and gear down to Atalaya with one night's stop in between. Erich, Jason and César were still a bit bleary in the morning, after a night at the town's ersatz disco, where our photographer passed out across one of the tables. We shoved off amidst whoops and hollers from a small crowd of well-wishers. The friendly townsfolk had been a welcome reprieve from the suspicious people we encountered upstream. As we cleared the mouth of the Perené and shot quickly down the Tambo the green of the jungle embraced us once again.

John and Pete paddled several hundred yards ahead while the rest of us drifted to check the map. They stayed out in front for a couple of hours until we finally reeled them in. We lost sight of them around bends in the river for almost an hour, though they knew this was a dangerous area. Erich's boat had yet to catch up, so we had no local guide or interpreter in case we ran into trouble.

When we finally caught up to Pete and John, I strongly suggested that we stick together for safety. Shortly thereafter, Erich's team motored up to join us and we all stopped at Catarata Koari, where a massive crystal-clear spring spilled out from underneath a cliff wall into the Tambo. We didn't have the time or energy to hike up the nearby canyon to see the large waterfall, but this break was well worth the stop. Team spirits were up, the skies were pleasantly overcast and the temperature was perfect.

Not long after, in the mist of low-flying mountain clouds, we came upon the Otica checkpoint. Erich's team arrived first and were busily showing papers and talking with the *fútbol*-jerseyed gunboys.

Checkpoint at Otica. Photo by Erich Schlegel

Each of the young guys carried the pervasive single-shot 20-gauge or .410 shotgun and sported soccer shoes with shin guards. They were enough to maintain order in the region but fell far short of the paramilitary firepower common in the jungle drug trade. A group of minivan-driving, yoga-pants-wearing, frappuccino-swilling soccer moms back home carried more heat than these boys. I suppose I should have been more reverent, but with the *alcalde*'s express written permission and the knowledge that their strongest firepower couldn't reach us in the middle of the river, I felt fairly comfortable.

The lookout tower built of roughhewn logs dominated the bluff between the river and the *fútbol* pitch, where the game had been interrupted by our presence. Everything went smoothly, though they wanted to go through every nook and cranny of my kayak, while leaving Zocimo's boat and the other two kayaks undisturbed. More curious than suspicious, a gunboy asked what action each little gadget performed. At first, he tried to grab my stuff from the hatches, until I stopped him and offered to show him myself, since I didn't want anything to get wet. When he asked about the red light flashing on my head-mounted video camera, I suddenly couldn't understand a word of Spanish and answered with the Third World Shrug noted by adventure travel writers Tim Cahill and P.J.

O'Rourke. A proper shrug is best performed with hands hung limply and palms facing forward. The eyebrows should always be raised to emphasize how stupid you are in your efforts to willfully not understand the simplest request. The goal isn't to convey a message, but merely to offer a non-response.

As I gave my best "I don't know nothing about no dinged-danged ol' video camera" act, Erich explained in perfect Spanish that it was, in fact, a video camera and West would be happy to turn it off so that you don't feel obliged to shoot us in the face, thank you. I ramped up my emphatic shrugging.

"¿Qué? ¿Qué? ¿Qué?" I repeated as Erich rolled his eyes and sternly directed me to turn off the stupid camera.

I looked at the gunboy, "¿Qué?"

Gunboy looked at Erich and rattled off something to the effect of "Tell this *pinche* gringo to turn off his stupid toy camera or I'll blast it off his head."

I looked at Erich, "¿Qué?"

"Turn off the damned camera."

I was on the verge of laughter when I thought it best to turn off the camera. Two minutes later, when they weren't looking, I turned it on again.

Gunboy assured us that he would call ahead to notify the *ronderos* in Poyeni of our arrival. In spite of the fact that he hadn't been contacted by our last checkpoint in Puerto Prado, we naively believed that this good man would call ahead.

The gorgeous river wound its way southeast through the steep green hills. Small farms and houses with wood slat walls and thatched roofs carved out their places in this paradise. Well-worn footpaths ran along the river, and fishermen tending their nets actually waved at us now and then, though more often our waves were met with the standard blank stare. At one point, I paddled near a woman on the right bank, dressed more like her Quechua cousins but without the bowler hat. For a minute or two she stood and stared, then took off running up the steep sandy bank as I drew closer.

Leonard Clark reported running several rapids on this short stretch of the Tambo River in 1946. Neither the rapids nor the narrow canyons he described in detail materialized for us, though the water flowed at a steady pace. Clark's account of his Amazon adventures tended to strain the bounds of credulity, as he wrote in detail of how he and his partners were taken prisoner by Campa headhunters, even though he had been granted witch-doctor status during an earlier ritual a few days upriver by the same individuals. Fortunately, he had learned the Campa language, translated the entire Bible into Campa (though the Campa didn't have written language) and escaped the headhunters all within the span of a week. His waterproof money belt and journal survived several swims through the rapids during his narrow escape.

While rafting down the Perené River, on his way to the Tambo, Clark's group was pursued by a rare Andean bear, which dropped its honeycomb and leapt into the raging river to pursue the raft, leaving Clark no choice but to shoot it right between the eyes at point blank range, just as it caught the raft. Unfortunately, piranhas stripped the bear down to its bones before the men could haul its carcass aboard the balsa raft. The same happened with a jaguar two days later.

If none of that was enough adventure for the week, Clark lassoed a 20-foot crocodile (the largest ever reported in the area, with an estimated weight of more than 1,000 lbs.), eventually driving a wooden post down its gullet and picking it up into the air to kill it. He then stripped naked and crept up on a herd of 1,000 wild boars, shooting one with an arrow from 80 feet away only to have the rest of the herd chase him up a tree. The horde of wild pigs was trying to knock over the tree when Clark's compadres managed to chase them away. Clark's accounts of similar exploits during his brief trip make for some very entertaining reading.

Ian and Jeff on the Rio Tambo. Photo by Erich Schlegel

Our trip through the same region was far less eventful. We encountered beautiful scenery and were held at gunpoint only twice, thus far.

For the most part, Pete and John continued well ahead of the rest of us, though with Zocimo's motorboat shadowing us, I was less concerned about our safety. The current was significant in the wave trains, allowing Jeff and Ian to hit their fastest sprint for the entire expedition on this stretch: 16 miles per hour. The rodeo ride through four-to-six-foot standing waves was quite exhilarating, leaving me grateful to have spent all that time in a whitewater kayak.

Small mountains gave way to hills and gravel bars became more common. I drifted for a bit to face upstream in order to video the magnificent landscape and sky. Jeff and Ian stayed with me while Pete and John continued on with Erich's boat. For 15 minutes the three of us waxed poetic about the rare place in which we found ourselves. How few people had the privilege of seeing what we were experiencing, right then and there. A large oddly angled cliff rose on river left, its unique rock layers appearing like a monolith laid askew. Rocks became fewer and fewer the further we traveled from the Andes proper into the jungle, so this huge outcropping was unique. We allowed the depth of the moment and the quiet sounds of the jungle to envelope us while we calmly drifted along, getting our money's worth from these short lives.

Around the next bend, on a high bluff, we found neat clapboard houses with sheet metal and thatched roofs amidst orderly rows of palm trees and wild St. Augustine grass. We paddled across the Tambo, now a quarter-mile wide, towards the shoreline just downriver of a rock pile where Zocimo's boat was beached alongside a couple of other riverboats. Shiny new corrugated metal sheets lay in disarray on the shore, ready for a new roof somewhere up the steep bluff. The predictable group of gunboys stood around looking simultaneously bored and curious. One kid wore his T-shirt over his head and face like a jungle ninja, but the others wore the standard uniform: shorts, bare feet, *fútbol* jersey, single-shot 20-gauge.

The entire team stood around staring at their feet like cowboys in a manicure shop. In my usual diplomatic way, I yelled out, "Howdy do!" as my kayak hit terra firma near Jason. He walked up to me quietly, without his usual smile.

"Things aren't good," he said under his breath. I took a quick tally and saw that no one in our group had been shot yet.

"What happened?"

I kept my tone down as I pulled my kayak ashore, ignoring the eyes of the gunboys. Jason explained that the jungle ninja fired a shot over the bow of Zocimo's boat when he didn't think Zocimo was pulling over fast enough. Zocimo was trying to avoid the upstream rocks and was clearly pointed towards shore and heading in, but jungle ninja boy was a bit trigger-happy. None of the other gunboys seemed alarmed. A call had been made to the *rondero* district manager, who was on his way.

Apparently, our friends in Otica had not called ahead to notify these *ronderos* of our impending arrival.

Jeff, Ian and I joined the toe-staring competition, during which we spoke in hushed tones to one another. This was the largest group of gunboys we'd seen, so Pete unnecessarily recommended we play it cool. I was all about the cool-playing, but also knew this was just business. Killing North Americans attracted all kinds of bad attention to those bent on quiet free-market capitalism.

Jason, César, West and Jeff making friends with the gunboys of Poyeni. Photo by Erich Schlegel

Before too long a river boat zoomed up with three guys aboard. While two gunboys secured the boat, the largest man we'd seen in a hundred miles disembarked and calmly walked toward us. He wasn't any taller than us, but compared to the skinny gunboys, he was hulking. A dark polo shirt stretched across his ample stomach and hung untucked over Bermuda shorts. On his feet were flip-flops and over his shoulder hung a 12-gauge pump, quite the upgrade from the squirrel guns carried by his minions. In his right hand, he held a light briefcase and on the tip of his nose balanced a small pair of reading glasses. Aside from the shotgun he looked like a banker on vacation. I doubt any of the gunboys would sweat a drop without his permission, nor would they hesitate if he told them to open fire.

As predicted, he was a businessman, and comported himself as such. This man had no doubt who he was or what he wanted, and he had no interest in trifling with minutia.

Erich and César explained our goal to the *jefe*, who nodded approvingly. Seemingly satisfied with our paperwork, the big man then turned his attention to Zocimo as Erich and César took a respectful step back. The conversation between the two men didn't result in the same warm fuzzy nods of approval. In fact, the only head movement I observed was Zocimo's noggin shaking in the universal sign of "nope, nope, nope, nope and nope." This wasn't going to end well for someone.

While we were on the good side of the *jefe*, Zocimo was on the naughty list for refusing to present his boat registration and river permit. To up the ante, he told the *jefe* that he didn't respect his authority and didn't have to show him no stinkin' papers. To this, the *rondero* remained calm but explained that he was the *only* authority in this area and it was his duty to make sure people had proper permits to conduct business, and that they were doing so in a boat that they rightfully owned. Zocimo called the *jefe* a thug and insisted he wouldn't show him the documents. Furthermore, Zocimo said that the *jefe* knew him and had seen him go up and down the river for years and he didn't deserve to be hassled like this.

The tide finally turned when the jungle banker calmly and quietly stated, "This boat is not going to go up this river or down this river unless I see your permits and boat registration." At that, Zocimo reluctantly showed his paperwork.

The *jefe* then gently explained to Erich and me that we were welcome in Poyeni, and that we and all of our gear would be safe under his authority. Then, this armed narco-trafficker left us with a word of advice: "You can never be too careful down here in the jungle. Many people, such as this boat captain, will take advantage of you or rob you. He should treat you much better. We are all brothers and must watch out for one another."

At that, he ordered the gunboys to treat us well, then shook our hands, wished us *buena suerte* and was off. The energetic gunboys worked like ants, carefully hauling our fully-loaded kayaks safely up the hill to a concrete pavilion with a sheet metal roof, where we set up camp.

Jason kept a close eye on the jungle ninja boy as he walked away from the group to pull the T-shirt off his head and put it back on his body. When he turned back towards us, he met Jason's stare. Jungle ninja boy tried to look away nonchalantly, but every time he glanced back, Jason was there. Turns out Jason didn't take well to getting shot at.

Once order was restored, we were welcome in Poyeni. Photo by Erich Schlegel

Jeff and Ian explaining what the hell we're doing to the boys in Poyeni. Photo by Erich Schlegel

A small contingent showed us through the neatly laid out dirt streets of Poyeni. The picturesque location was home to several hundred inhabitants with a proper Plaza de Armas, glass-windowed community center and many *tiendas*. Street lights remained dark, though generators fired up here and there. Our tour guides explained that tourists usually slept in the civic center during their stay, which begged the question: Are all tourists met at gunpoint on arrival? We were safe and sound, but Poyeni really did need to soften its welcome committee.

On a concrete tabletop we laid the road map of Peru, which was our primary form of navigation down the Amazon, together with the glossy tourist map we were given in Puerto Prado. It felt good to see significant progress in just a few days, compared to the slow pace through the whitewater canyons of the Mantaro. A colorful Buick-sized bug landed on the map to give his opinion, startling the gringos. César gently hoisted the beast for closer inspection, his muscles straining to hold the six-legged monster aloft. Hell, I'd seen smaller livestock. It was a beautiful yellowish orange with a pleasantly calm demeanor, which was fortunate since we couldn't find a club big enough to do anything but piss it off. César chuckled at our misgivings as I silently wished the bug would eat one of his fingers.

Photo by Erich Schlegel

Morning on the Tambo. Photo by Erich Schlegel

Chapter Eighteen
Atalaya

Day 49, Mile 737

"I look at victory as milestones on a very long highway."
Joan Benoit

After being unceremoniously shooed away from our riverside pavilion by the pre-dawn rush of women setting up to sell their wares, we hit the river early for the short and uneventful trip to Atalaya. Signs of modernity became more common with the approach to town, where the Urubamba and Tambo river valleys flattened into distant verdant hills and motorboat traffic increased.

Over the past century, Atalaya has become a geographic milestone for Amazon River expeditions. The small outpost is where J. Calvin Giddings and his team ended the first descent of the Apurímac in 1975, with several members refusing to speak to one another. In *Demon River Apurímac*, his thrilling book on the groundbreaking expedition, Giddings describes the teammates glowering at one another across a muddy *fútbol* pitch as two local teams played.

Expeditions are hard on team cohesion. The unique combination of constant danger, hunger, fatigue and uncertainty tests even the strongest partnerships. A decade after Giddings' first descent, the stress of running the Apurímac fractured the team that Francois Odendaal assembled to complete the first source-to-sea descent of the Amazon. This was the expedition Joe Kane and Piotr joined, and Odendaal's leadership crumbled under the pressure. The conflict came to a head when Odendaal tried to remove Kane from the expedition. When the team resisted Kane's dismissal, Odendaal demanded a vote of confidence in his leadership, and lost. "He hadn't done his sums," team member Jerome Truran said. Piotr became the expedition's de facto leader and continued to the Atlantic with Kane, who wrote the truly amazing bestseller *Running the Amazon* about the fractured, but ultimately successful, expedition. I read Kane's book enough times to be wary of team dynamics, especially on this part of the river.

As we were packing the last bit of gear and kayaks into our Atalaya hostel storage area, a gregarious fellow pulled up to the curb on a motorcycle with a woman and young girl balanced on the back of the seat. Miguel Hernandez introduced himself to Jason with a flourish, and then launched into a barrage of Spanish far faster than we could follow. Eventually, Miguel produced a card, indicating he was in the radio business. Jason took this as a sign that Miguel would be a good buck to pass along to Erich, who specialized in fast-talking media types. Erich learned that Miguel was a popular radio disc jockey, with a show specializing in all the pertinent jungle news—a kind of "Jungle Home Companion," with a cheesy disco soundtrack and messages to friends, family, law enforcement, narco-traffickers, illegal loggers and gold diggers.

At Miguel's suggestion, we all shuffled down to a restaurant on the promenade, along with his wife and daughter. The huge speakers in the courtyard of the large café vomited massive amounts of hyper-bass dance beat electronica for the entertainment of us and the only other patrons, an elderly couple sitting in stunned silence.

To kick off a familiar comic ritual, I asked the waiter to lower the volume. The kid asked me to repeat myself, as he was having trouble hearing my request over the loud music. We recapped this Abbott and Costello routine several times until finally he got the message, whereupon he stepped over to the bartender and began the same comic exchange all over again. The bartender eventually reached behind the bar and lowered the volume an infinitesimal amount. This same routine occurs in faux tropical bars throughout the word. Perhaps the owners think everyone within earshot will break out dancing and throw money all over the floor. Eventually, I convinced the bartender to shove a handful of money into the sound system to quiet it down and gave him the appropriate wad to accomplish the job.

Miguel knew everyone and was full of promises to introduce us to the mayor, get us a support boat and military escort, and yadda, yadda, yadda. Miguel ordered round after round of beer and food, and when the dinner began to wind down he ordered more food to go. When the bill came, Miguel regarded it with the same forced indifference subway workers reserve for the third rail.

The café walls were covered in paintings and sculptures of mermaids and muscular men straddling, writhing, humping, slithering and grunting in what looked like sexual positions, or perhaps interpretive dance if having sex was what they were trying to interpret. Naked buxom women cast in stone with rapturous expressions grasped serpents, vines and dolphins in steamy jungle settings. The artist definitely had a targeted imagination and perhaps needed a night out on the town.

For the first time since entering the Red Zone, I dropped my guard a bit. Adding to the miasma of emotions that come with having guns pointed at us, plus substandard coffee, John's creeping disappointment and the milestone of reaching Atalaya, it was also my wife's birthday—the first I had missed in our 24 years of dating and marriage. Knowing this would be the case, I had given my daughter Isabella some money and arranged for Jeff's girlfriend to take her shopping for some jewelry and dinner to surprise Lizet in my absence. I really wanted Lizet to know I missed her and had planned well ahead of time for her birthday. I thought about my girls all the time ,especially today. Being apart on this special day seemed to double the weight on my shoulders.

In a rare quiet moment early in the evening, I went out to the pretty concrete promenade overlooking the Ucayali to make a satellite phone call to wish Lizet a happy birthday. I asked about the events of the evening, only to find out that Isabella had asked to spend time with her friends instead of taking her mom out for a nice birthday dinner with the money I left her. It was a typical move for a 13-year-old girl and I should have let it go, as Lizet really didn't mind. Instead I asked Lizet to pass the phone and proceeded to read my daughter the riot act from 2,000 miles away. She ended up in tears and I ended the call feeling like crap. I missed my girls and screwed up the call, completely. I was a horrible dad. Ian saw that I was upset and asked what happened. I told him, and in a simple manner that I came to respect he said: "Well, that could have been handled better."

That Night in Atalaya

> *"Welcome to the jungle we got fun and games*
> *We got everything you want honey, we know the names*
> *We are the people that can find whatever you may need*
> *If you got the money honey, we got your disease"*
>
> William Bruce Rose, Jr.

At the ass end of the Red Zone, Atalaya is one of the rare jungle outposts accessible overland from the outside world, though the route requires a slow crawl through mountains choked with mud and vegetation bent on reclaiming its rightful space. As usual, we stood out like pale lumbering aliens as we strolled through town.

From this point on Miguel would be inseparable from, and intermittently helpful to, the expedition all the way to Pucallpa, about 350 miles downstream. True to his word, he interviewed Erich at the radio station and introduced Erich, Jason and César to the mayor. They received a personal tour of the town, surrounding area and the mayor's home. By the time we departed the next day, Miguel had arranged for Erich to use the mayor's speed boat to transport his team and gear to Pucallpa, merely for the price of gas.

As the paddlers settled in for a rare night of showers and sheets, Erich, Jason and César fulfilled their public relations duties, paying for Miguel's drinks and tips in one raucous disco strip bar after another. The team truly needed to let off a little steam, after enduring guns and other dangers. Hell, just living day-to-day in our situations was stressful. The legend that follows is pieced together from the survivors, with each depiction differing in significant details, though each telling began with some variation of "There I was, just minding my own business…"

The sweaty, crowded bar shook with the hyper-bass and torturous electronica beat dialed up to 11. Talk was impossible, requiring the manly patronage to scream at one another. Quart bottles of beer flowed through kidneys with amazing velocity, depositing necessary levels of alcohol along the way and raising the level of the Ucayali River from rivulets along the outside wall of the establishment. Members of the fairer sex were in a noticeable minority, with the scant few in attendance purely profiting from primal male stupidity and libidinal despair.

In the center of the sodden testosterone mosh pit was a dancer who appeared to be a popular entertainer for the establishment. Crawling onto the stage, she wasted no time stripping down to pretty much nothing, all the while collecting *soles* and whipping the crowd into a drunken, sweaty frenzy. At the urging of the DJ, she picked Erich as her target for the night. The rest of the group, always looking after the health of the team, made futile attempts to keep our intrepid photographer planted on terra firma, but they were no match for the writhing jungle mistress. Jumping down into the melee, with the crowd whooping and hollering, she dragged our mildly hesitant, severely impaired, teammate onto the sparkling stage. She then commenced to bump, grind and otherwise slither every available body part all over Erich to the techno-disco beat that drowned out the volcanic crowd, now locked in a monosyllabic, grunting, knuckle-dragging, yelling mass of stinking man-dom. Under the direction of the DJ's verbal choreography, the stripper lost her final scraps of clothing. It was hard to believe the crowd could go more feverish, but on they barreled, with pedal to the metal in an out of control bus with bald tires and flaming brakes all the way to crazy town.

Just getting warmed up, the naked artiste blindfolded Erich and relieved him of his burdensome clothing. With all the dance rhythm a play-that-funky-music white boy could muster, our Texan boogied like a grinning rhino doing the Macarena on a diving board covered in syrup. Now, there aren't a lot of wallflowers in the Texas canoe racing community, and Erich leaned precipitously toward the more gregarious end of that bell curve.

It's at this point the accounts of the drunken survivors' part, with some reporting that our valiant explorer kept himself securely holstered, while others clearly saw the flag at full mast, waving proudly. Granted, there's evidence aplenty that all involved may have been a bit impaired.

After failing to apply a condom, in spite of numerous attempts, and desperate to please the screaming horde, the stripper poured cold beer over our hero's head, calming the savage beast back into submission long enough for him to get dressed and recede back into the frothing mob. Suffice it to say, our champion left with his virtue threadbare, but still firmly intact.

The remainder of the night, the boys report a thick humid blur of racing around town, sweaty women, red lights, more beer, more rum, more women and more dancing. At one seedy bar a small group of locals spotted the sore thumbs in the crowd and immediately tried to blend in with the walls. It was this very group that held us at gunpoint the day before in Poyeni and worse yet, the kid that took a potshot over the bow with his scattergun. Though his senses were chemically dulled, his memory was keen and it didn't take much to bring Jason to stone cold sober as soon as he picked out the gunboy who took the shot at Zocimo's boat. Jason locked eyes with the gunboy and brought him to the Erich's attention, announcing his intention to beat the holy crap out of the kid and anyone the kid wanted to bring along for an ass whupping. In a surprising and fleeting fit of lucidity, César, Manuel and Erich grabbed Jason just as he launched out of his chair to single-handedly assault the entire *rondero* crew, and to Jason's credit the posse of gunboys soberly retreated en masse.

As the saying goes, never pick a fight with an old guy: I mean for heaven's sake, he's sore, injured, achy and gets tired easily, so he'll probably just kill you. Under Erich's inebriated guidance, the group moved on to another bar to administer more medicinal rum, dragging Jason away from an all-out attack on the gunboys, who were substantially less bold without their firearms.

A scant few hours later, Jason drew the short straw at daybreak, appearing slack jawed, quite the whiter shade of pale, listing heavily to starboard and expressionless behind his sunglasses, and acted as spokesman for the trio of lost boys in the morning glare. I swear, his skin emitted puffs of smoke as he shuffled painfully from the shadow of the hotel portico to the sunny sidewalk. I was afraid he'd spontaneously combust. Jason mumbled their plan to meet us down river later that day once they secured the mayor's boat and healed up a bit. The mayor came down to see us off and have the local paper take pictures of him sitting in my kayak on the sidewalk.

On the muddy shore, amidst the colorful riverboats and pecky peckys, I took a closer look at the astounding quantity of gear John and Pete stuffed into their tandem kayak. Here I was concerned with taking an extra tube of toothpaste, while Pete shoved a massive tool kit into his kayak, including tools we already had in the other boats, and some that had no conceivable use on a kayak. Perhaps he figured we'd come across a John Deere tractor that needed rebuilding. I told Pete it might be a good idea for he and John to leave some stuff with Erich.

"We're fine," he said.

But they weren't fine. Not by a long shot.

Chapter Nineteen
Death and Dismemberment on the Ucayali

Day 50, Mile 792

> *"It was written I should be loyal to the nightmare of my choice."*
> Joseph Conrad

As we left Atalaya, the fate of a Polish couple murdered the previous year on this stretch of river weighed heavily on all our minds. Jaroslaw Frackiewicz, 70, and his wife Celina Mroz, 58, left Atalaya on May 25, 2011 in a tandem kayak. Both were experienced kayakers, having paddled extensively in Canada, India and Egypt, and they kept an online blog about their Peruvian adventure. The killers may never have been apprehended if friends and relatives in Poland hadn't contacted local authorities weeks after the blog went dark.

Peruvian investigators eventually tracked down one of the assailants, who confessed to everything and fingered his uncle and brother-in-law as the triggermen. He told police that as the couple approached in their kayak, one of the killers shot Jaroslaw in the chest without so much as a word, wounding him. Then five men, drunk on *masato* (cassava plant chewed, spit and fermented) chased down the couple in a canoe. They came alongside and, again without speaking, unleashed a barrage of shotgun fire and arrows at little more than arm's length.

They then dismembered the bodies, weighted down the remains with rocks and sank them into the depths of the river, according to witnesses who saw the murderers paddling the couple's kayak. Investigators later found some of the couple's belongings in one of the assailant's homes, and three of the men were convicted and jailed. The other two remain at large, presumably armed, nearby and harboring hostile intent toward *pela caras* in kayaks. The Poles had been ambushed one day after leaving Atalaya. Given our faster pace, we would pass the scene of their murder in a matter of hours.

The outskirts of Atalaya fell away as we paddled the last miles of the Tambo River to its confluence with the Urubamba, where it becomes the Ucayali River—our last and longest in the succession of tributaries carrying us to the Amazon proper, some 900 miles downstream. With the added flow from the Urubamba we expected the current to increase substantially, but it didn't provide the added push we anticipated. The confluence did bring increased river traffic, however. Below the confluence, the river was full of barges loaded with everything from logs and construction equipment to bunches of bananas and crates of soft drinks. Passengers and crew regarded us with blank stares. Waving, apparently, was as foreign to them as chicken fried steak.

Logging on the Ucayali. Photo by Erich Schlegel

We heard our first logging camp well before we saw it. Techno dance music wafted upriver, led by a thunderously unnatural bass beat. Rounding a bend, we found huge tree trunks stacked high on the bank, ready for loading onto a barge. A bright yellow front-end loader pushed and hoisted the massive logs as men moved about in a clearing carved about 100 yards into the dense jungle. The workers returned our smiles and waves. Our people.

As we traveled deeper into the jungle it struck me how starkly out-of-place the objects of western industry had become when placed in the folds of the jungle. The unnaturally bright yellow Caterpillar loader against the backdrop of ultra-lush, greener-than-green forest was like a mustard stain on a starched dress shirt.

When we paddled close to the bank the jungle seemed to whip by, as the water hissed loudly through downed trees and foliage. These strainers could trap and drown an unwary kayaker, so we gave them a wide berth. We made good time on the Ucayali, though far from the 10 mph average we'd expected to maintain. The tiny GPS mounted on my kayak's front deck dutifully reported our speed and distance. After five days of flatwater and only 270 miles, it was clear that there was no way we were going to average 100 miles per day as I planned. I decided not to push it. I was still worn from the whitewater and we had far too much river in front of us to hit the afterburners this early on. Besides, the team had yet to settle into a steady rhythm, which is a hallmark of all successful expeditions.

I thought a great deal about how to bring John into the fold. To me the expedition was a rare indulgence, and we were among the fortunate few to paddle the greatest river on earth, which all seemed lost on John. All we heard from him was complaints, and everything I heard about him was complaints. We were in the midst of a magnificent eight-course meal in the world's greatest restaurant, and John didn't like the pattern of the silverware. I just couldn't figure out why he was so disappointed, and he wasn't talking.

Just as the sun set below the rainforest, Erich and company pulled up in a good-sized motorboat that resembled a small yellow spaceship. The mayor's boat was a covered sheet metal rig with a large outboard motor. Manning the wheel was Captain Rambo, whose affect was that of a man enjoying every second of a fabulous life. Miguel managed to invite himself along, though his services were a bit of a mystery to everyone, especially when physical effort seemed imminent. Raising a tent, gathering wood or carrying gear was out of the question, owing to his bad back. He had no problem opening a bottle, though opening his wallet seemed far too strenuous a task. Whenever that activity was called for he blended into the background, allowing Erich or Jason to handle the heavy labor.

Before hitting the sack, I called Isabella on the sat phone. It was an indulgent, expensive, unscheduled call, but I figured it to be pretty important. I apologized for my childish behavior the night before on the phone and told her how much I missed her. If ever a parent can provide an example for a single character trait into their child, I recommend contrition.

Rambo, king of the fastboats. Photo by Erich Schlegel

The Devil's Bend

> *"This strait is the first serious difficulty encountered in ascending the Ucayali; the current dashes with much violence against the trunks of large trees which lodge in, and almost block up, the passage."*
>
> John Randolph Tucker

The next day the river split around a series of large islands, and the left-hand channel makes a long sweeping turn known as the Vuelta del Diablo. The name dates to at least 1873 when John Randolph Tucker, a distinguished veteran of the U.S. and Confederate navies, then serving as a Rear Admiral in the Peruvian service, charted this particularly treacherous bend. He called it the "The Devil's Leap" and noted it in his official report as the only hindrance to steamships making their way up the Ucayali to trade on the Tambo.

The obvious route was to stay right and avoid the Diablo but, as so often happens on the Amazon, things were not as they first appeared. We drifted while determining the correct course, and by the time we'd consulted the map and GPS to determine that we absolutely needed to avoid the left channel, it was too late. All three kayaks were in the grip of the strong left-hand current. We dug in hard to inch back up against the strong flow to the split so we could take the much shorter and safer route to the right (east) of the islands. Jeff and Ian made short work of it swinging single blades. I had to put out more effort but managed to hang with them. Pete and John flung huge buckets with their double blades in asynchrony, but barely managed to keep even with the current they were fighting.

To Ian and Jeff, I called out a change of plan: we'd take the left channel. Ian smiled and quietly commended me for my transparent diplomacy. It took us quite a while to get around that group of islands, which was the first thing Erich's crew noted several hours later at the confluence below the islands. They killed time pretending to fish and complaining about our poor navigation. They were glad to see us, but quickly zoomed ahead to the town of Bolognesi a few miles downstream.

Soon after they disappeared around the bend, Pete inquired, "Is this the area where that adventure tuber was shot?" It was, in fact, the very place where Davey DuPlessis had been attacked.

Erich, the voice of the expedition over the Amazon airwaves. Photo by Jason Jones

Erich got pulled into a radio interview in Bolognesi, so we pushed past. We finally That night Erich shared the news that Joe Kane wanted to meet in Iquitos for the National Geographic interview. Everything seemed to be on track, and César said the river would stay wide and slow all the way to Pucallpa, after which the current would increase a bit. When the Ucayali joined the Marañón, it would really pick up speed. Until then, we'd be in the horse latitudes. My shoulder was feeling much better since I'd adjusted my stroke to a less efficient, but less painful technique. Earlier in the day, I tried surfing the wake of a tugboat but a sharp pinch reminded me not to press my luck. Just to be safe, I administered an extra dose of rum.

The threat of rain persuaded me to set the rainfly, which turned my tent into a perfect sauna. The camp seemed crowded, leaving me a bit irritable, which I kept to myself. Perhaps I was feeling the burden of all this extra gear and activity when all I really wanted was to drift unhindered down the Amazon. It was not my intention to have Erich hauling so much of our gear, and I certainly hadn't expected John's surly disposition. Miguel—still hanging on—was a third wheel whose presence I didn't welcome, but he was Erich's issue. All the other guys were great and I didn't want my dour mood to spill over onto them, so I hit the sack early with rivulets of sweat streaming down my skin, listening to the sound of curious dolphins huffing in the dark.

An illegal logging operation drifting downstream to a mill. Photo by Erich Schlegel

Pete and West survey the digs in Galilea. Photo by Erich Schlegel

Galilea

Day 53, Mile 990

"The nice thing about living in a small town is that when you don't know what you're doing, someone else does."

Emmanuel Kant

At Galilea, the bow of my pristine white sea kayak plowed into a literal shithole, a stagnant eddy along a steep riverbank lined with open-air structures and rickety outhouses oozing their effluent straight down the bank to our landing spot. We stood ankle deep in the slippery scum as the sun sank over the endless green leafy ocean, behind the buildings crowding the river's edge. The day's effort had been a tepid sluggish flow reminiscent of a death march through mud. Under an overcast sky, all we could do was keep trudging through thick, stagnant air and placid water.

Main street Galilea was a markedly pleasant dirt plaza with wooden benches in the middle. Each wood-slat building had its own generator pock-pocking out just enough juice for a few dim bulbs and a refrigerator. Generators were shut down late at night to allow spoilable foodstuffs to fend for themselves. Late the next day, the asthmatic generators were begrudgingly called into action once again to catch whatever bacteria threatened to turn the food into yet another poop-liquefying entrée. It was a lesson in economics and pragmatism, pitting the cost of gasoline against the reliability of the cheap Chinese generators and the ever-present bacteria in food rinsed in untreated water. I took my fish well-done, thank you.

Nightlife and cafe in Galilea. Photo by Erich Schlegel

We were a couple of dozen miles from the place where Juliane Koepcke crash landed, after the airliner in which she and her mother were flying broke up during a fierce thunderstorm on Christmas Eve, 1971 with 92 souls aboard. Juliane, then a 17-year-old daughter of Amazon jungle researchers, was the only survivor. She awoke on the ground, still strapped in her chair after a one-mile freefall, and made her way back to safety after 11 days in the jungle. Her injuries included a broken clavicle, swollen eye and cut on her arm. During her jungle ordeal, she couldn't sleep, due to the insects and an infection from numerous botflies. Eventually, she found a camp and extracted the botfly larva with gasoline. Hunters eventually found her and delivered her to safety.

Permeating the wafts of diesel, gasoline, raw sewage and roasted vertebrate was the moist lush aroma of the jungle. In spite of whatever mankind did to mask the jungle, the scent of fresh life was omnipresent, an unstoppable force reclaiming ground with patient strength. As in Gabriel Garcia Marquez's *100 Years of Solitude*, it took only a scant bit of time, a mere blip on the eternal calendar, for nature to grow, engulf, inundate, embrace and devour whatever man thought was strong and permanent. The vibrant smell of the jungle leaned heavily on the flimsy door of Galilea, daring townsfolk to leave it the tiniest bit ajar. The wondrous smell was its own omen, a reminder to me and everyone in Galilea that we were but itinerant renters of nature's home. All will eventually be covered and forgotten in a flash of green.

Pete spread out his sleeping bag near the boats, as was his preference, while Rambo and César stayed on the fastboat. The rest of us got rooms at the *hospedaje*.

The two-story refuge had walls of vertical planks, with quarter-inch gaps between them to allow easy access for mosquitoes. An unmuffled diesel generator grunted every half-minute or so, laboring to supply barely enough juice to the bare bulbs that barely hung by bare wires in the bare rooms. A single sheet covered the saggy stained mattresses sitting off the floor on lumber frames.

The disco stood at one end of the 100-yard-long settlement, announcing its presence with a brightly painted sign and incessant hyper-bass 1980s techno pop rehash doing battle with the establishment's own pocking generators. At the other end of the town was a row of *tiendas*, each with open-air tables for dining. Neat shelves in each kiosk held the exact same products as its neighbors, not 20 feet away. Every café served the same bony fish, chicken, rice and plantains, all of which had been freshly killed, caught and/or cooked within the last week or so. A bottle of cheap rum helped sterilize each bite as we cleaned our plates in an effort to replenish our dwindling energy reserves. Galileans were out in the square, visiting one another, playing cards and pick-up *fútbol*, flirting and socializing as people all over the world have done for centuries, except in present-day America where we sequester ourselves inside safe, hermetically sealed houses to watch people on television pretending to have lives more interesting than ours. It was a wonderful, lively atmosphere in Galilea, which picked up our spirits dulled by the day's humid death march.

While our group garnered some attention, Galileans didn't tend to stand around gawking at us, which was a welcome change. Shopkeepers were friendly and talkative, questioning our expedition and more than happy to answer questions about their lives in the jungle.

After supper, Pete confided to Jeff, Ian and me that the past week in the boat with John had been "pure hell." He didn't provide details, but confirmed that John was generally dissatisfied with decisions I made as the expedition's leader, among other issues. By this time, John had established a habit of eating apart from the rest of us whenever we stopped for lunch and kept to himself at camp. He spoke only to Pete unless someone else addressed him directly. John kept his own stash of food, though everyone pulled from a communal storage system and tended to socialize around meals.

I brought up the idea of replacing John with another paddler. Given that I paid for everyone's airfare, hotel rooms, food, gear and the tandem kayaks they paddled, the buck stopped with me, but for such an important decision I wasn't going to act without the support of the team. Thus far, I'd been able to somewhat ignore John's recalcitrance. The expedition was still moving toward the Atlantic, and while John wasn't hindering our progress he certainly wasn't contributing to the team effort.

I wanted to give John a chance to voice his concerns and to hear the concerns of the team, though Jeff and Pete felt strongly that John had a cowboy outlook and it was futile to expect him to change or talk about issues. Each of us at that table had run cattle. We all owned actual work trucks and guns and had spent decades working construction. The idea that John wasn't going to talk or start working as a team member simply because of some contrived cowboy stoicism didn't seem valid. Ian and I believed it was important that I try to work things out before simply replacing John. We left the impromptu meeting with a plan for me to talk with John in Pucallpa, a long day's journey downstream.

Our human insignificance manifested once the generators were silenced and the insect chorus enfolded every available cubic inch of air around us. Texas cicadas have always brought a warm, homey melancholy to evenings of my youth, but the roar of a million species here in the Amazon was something completely different. It was quite deafening but not altogether unwelcome.

In the middle of the night I was awakened once again by a foul gut. Drenched in sweat, I stumbled down the darkened uneven loose plank stairs to find the outhouse, which was behind the *hospedaje* somewhere. In my delirious state, I finally discovered the lean-to, but couldn't conjure any relief. The super bright moon lit up Galilea like a mid-day sun, so I lurched to the benches in the vacant town center and sprawled out naked to cool down from my oven-like room. After an hour of sipping water from my Nalgene bottle, I recovered from the heat exhaustion. At four a.m., we were back on the water for the long day push to Pucallpa, but not before Jeff met his roommate for the night, as he packed to leave. While bent over his bag, a large rat leapt from a tall shelf, bounced off his back then onto the bed before scurrying underneath.

The Cavalry

By mid-afternoon, we reached the mouth of the Río Pachitea. In 1866, a steamer under the command of a Captain Vargas, chugged up the Pachitea and from there another 60 miles further up the Río Palcazu in search of a water route into the Andean highlands. This is the farthest any expedition had ascended the Río Ucayali tributary, thus far. Two officers, Tavara and West, went ashore to ask the local people where they might find a good Thai food joint or Starbucks, and their kind howdy-do was returned with a quick and painful death. Captain Vargas and company bid a hasty retreat.

Another expedition was launched to punish the local people for their poor manners. Led by the governor of the region, the young and eager Don Benito Arana, military types loaded onto three steamers, found the location of the slayings and massacred a good number of locals before they burned their village to make sure the message was clear. Following the Palcazu and then the Río Mairo as far up as the water depth allowed, Arana stuck a flag in the ground and shrewdly proclaimed the place Puerto Prado, after the Peruvian president at the time. Arana proudly reported that the juncture was a mere 10-day trek over the Andes from Lima, and 12 days by river to Iquitos and the Amazon proper. A cross-continent trade route seemed within reach. Unfortunately for the enthusiastic young governor, the waters of the Palcazu and Pachitea returned to unnavigable low levels within days of his retreat.

A few months later, Admiral Tucker, accompanied by Dr. Santiago Tavara—the brother of one of the slain officers—launched canoes from the proposed site of Puerto Prado on the Palcazu and reached the Ucayali within a few days to meet the steamer *Morona*. While Tucker supported the concept of a commercial route via the Palcazu, Dr. Tavara opposed the route due to erratic water levels. He favored a route which came down from Tarma to the Perené and spilled into the Río Tambo, near present-day Puerto Prado.

We arrived at the mouth of the Pachitea, with pecky peckys rattling here and there, river taxis coming and going and fisherman milling around makeshift lean-tos on both banks. The Ucayali widened significantly and the floodplain, indicated by the tree line far away from the river, was easily several miles across. We had a scare when a fastboat veered from its course and charged directly toward our little group of kayaks. The pilot slowed as he buzzed passed, revealing a brightly painted boat with the name of a tourist company in huge letters stenciled on its side. The smiling, waving passengers wore bright orange horse-collar life jackets. Given their western clothing and girth, these were tourists from afar. After a moment of picture taking and mobile phones held aloft for video, they were off.

This incident, though a false alarm, was a reminder that we were heading into a known pirate area. Up until now, we hadn't really practiced our survival plan, but I thought it was time. I handed out walkie-talkies to the respective kayaks, then we paddled a few hundred yards away from one another to test them out.

"Jeff, can you hear me?" I spoke into the little handheld.

"Roger that. Can you hear me? Over." He came through clear as day.

"Yepper. Over," I replied.

"John, can you hear me and Jeff? Over."

After hearing no response, I pushed the button and repeated the message. No response, so I repeated the message two or three more times. No response.

"Jeff, are you hearing my calls to John? Over."

"Loud and clear. Over," came his response.

The three kayaks converged. I asked John if his walkie-talkie was on. He told me it was and that he heard me. He reached down to his feet and held up the walkie-talkie, which was still in the Ziploc baggie.

"Why didn't you answer?"

"I did answer," he responded.

"How? Telepathically?" I was incredulous that he had made such a simple thing so difficult.

"I told Pete." John answered matter-of-factly, as if this were a reasonable thing to do.

"Yeah," Pete offered. "I heard it." Jeff and Ian laughed.

"Oh, for Christ's sake, why the hell can't you just work with us?" I paddled off, adding another straw to the proverbial camel's back.

Just about then, another speedboat approached. Uh-oh. I hadn't expected to put our plan to separate into action so soon. Erich's boat was right on it, zooming up to come between the kayaks and the speedboat. I was alarmed to see several guns silhouetted against the dull glare of the huge sky. Crap.

Only when they got within 50 yards or so were we able to make out the jungle fatigues and military caps worn by the men crowding the large open-air boat. The cavalry had arrived. I figured their early presence wasn't a good sign. The captain conversed with César and Erich for a few minutes, then we all met at the nearest mud bank for proper introductions.

John and Pete under the watchful eye of the Peruvian Coast Guard. Photo by Erich Schlegel

Over cups of camp stove coffee, we heard about the pirate activity around the Río Pachitea river mouth, with frequent armed hit-and-run attacks reported throughout the area. The usual method was for the pirates to zoom up, point guns and take what they want. If anyone objected, even a little, they'd be shot. The military began patrolling the area the previous year, so they really didn't have to go out of their way to keep an eye on us as we pushed into Pucallpa. Following introductions, small talk and pictures, we were back on our way with a substantial bit of stress lifted off our shoulders.

The high-powered Coast Guard boat shadowed our pokey kayaks the remainder of the day. After an hour or so, Jason yelled over to me, "Man, with all this firepower I almost wish someone would start some shit!" He'd been itching for some action ever since the punk took a shot at his boat back in Poyeni. In spite of his Hawaiian shirt and subpar Spanish, the Peruvian servicemen regarded Jason as a brother-in-arms. More than once a heavily shrouded ratty looking pecky-pecky veered towards our kayaks, only to incur the roar of the twin outboards from our escort and a battery of automatic weaponry pointed in their direction. We were the wimpy kid in the schoolyard who just made friends with the biggest kid on the block.

Jason and his Peruvian brethren. Photo by Erich Schlegel

During our lunch break, we offered the uniforms some of our food, but after a look at our cans of tuna and crackers, they politely declined. Far from the inconvenience I assumed we presented to the Coast Guard, our expedition was an enjoyable diversion from their usual routine.

At a split in the river around a large island the Coast Guard went right, and Jeff, Ian and I paddled from the left bank a mile across the river to follow our protectors. John and Pete were about 30 yards behind us when we started to cross, and it wasn't until we were almost all the way across that we discovered they were taking the smaller left channel around the island. There was no way to tell whether the course met up below the island or went up into the jungle or just came to a dead stop in some lagoon. Our two compadres left us, like expendable extras in a B-grade horror movie, the ones who think it's a great idea to split up in order to more effectively seek out the source of the blood curdling scream. I called out to Jeff and Ian, who were just as stumped by their move. My confusion quickly turned to irritation at the potential danger John and Pete brought upon the expedition. Within minutes they were out of sight. Our escorts searched the horizon with binoculars in vain for the tandem kayak, but it disappeared behind the island.

Things quickly got awkward. The Coast Guard captain was upset, but professional, voicing his concerns about John and Pete separating in this dangerous area. I was embarrassed and promised it wouldn't happen again. Using the walkie-talkie, I called out for our teammates, though I knew it was futile. After some very tense waiting, the kayak came through a small shallow slough into a lagoon where we were all waiting.

With the best upbeat "hey let's all get along" voice I could muster, I told Pete and John that we needed to stay together because it was such a dangerous area. John responded that he didn't have a GPS, therefore he didn't have the ability to tell the best route to take. I told him that we were all following the Coast Guard, therefore a GPS was irrelevant. They could have easily followed us when we crossed the river before the split.

Toward day's end, I was worn out and falling behind. At one point, to get ahead and rest for a while, I caught an oddly fast current inside a large open bend and cut across as the rest of the team went around the outside. I kept an eye on the group over my right shoulder and waited for them where the river narrowed again. Jeff was pissed off at my move, and rightly so. I really hadn't considered it a big deal, since we were all in sight of one another the entire time but then realized he was right. From then on, I vowed to stick with them and talk about any changes in course. They agreed to slow down for me when I was worn out.

As the sun set, we donned our head lights, which weren't much help in seeing things on the huge river but helped other boats to see us. A steady line of pecky peckys of various sizes and all dangerously close to sinking, made their way downstream along the right bank—the Amazon's version of rush hour. Rounding a huge bend to the right, we were shocked by the sudden appearance of a large well-lit industrial city. Pucallpa was awash with multi-story office buildings of various age, neon lights, street lamps, traffic and the loud hum of modernism along a harbor lined with tall tug boats, cranes, various small ships and massive barges.

Erich's team had gone ahead hours before to secure accommodations and a place to store the kayaks for the night. Via Barbara, they told us to go to a three-story tall hospital barge run by the Coast Guard. Invisible and barely visible small craft traffic on the wide black river increased to a frenetic level, zipping to and fro. Industrial noises filled the darkness against the blinding lights, and out of the blackness now and then an unmuffled pecky-pecky barely missed whichever kayak was closest, then disappeared like a loud clumsy bat. The Coast Guard boat hung close, shining their sun-bright spotlight ahead, sweeping the darkness for floating debris, which we did our best to avoid.

White knuckled in the melee, the kayaks stuck close together. At one point, with my kayak about 10 feet to the left of John and Pete, Jeff and Ian crept up between us to ride our wakes, a common technique in marathon canoeing, similar to drafting in a bike race. As soon as Jeff steered into position, John hit a hard-left rudder pedal, closing the gap between our two boats. I responded with an evasive turn of my own, thinking John was swerving to avoid some debris in the dark river. Pete called out a startled, "Whoa! Whoa!" to John.

Jeff and Ian backed off, then returned to attain the draft position once again. This time, John hit another hard-left rudder, almost ramming Jeff's kayak. Pete called out again, "Whoa! Whoa!" as Jeff quickly swerved to avoid the crash.

"You getting the message now!" John yelled angrily.

"What the hell is all that about!" I shouted, but John didn't respond. Ian and I exchanged bewildered looks, though I later learned that Jeff—with his somewhat numb hearing—hadn't heard John hollering at him. Jeff moved his boat to the other side of me, so we were all paddling abreast. A few minutes later we found Erich and company at our port of call at the water's edge of the jungle metropolis.

Chapter Twenty
The Pucallpa Accord

Day 54, Mile 1,071

"Out beyond the ideas of right doing and wrong doing, there is a field. I will meet you there."

Rumi

Pucallpa is a busy contemporary city that keeps a feverish pace, especially by night. Pucallpans make the most of it, with each night an excuse for a city-wide party. We launched headlong, dirty and smelly, into the melee with a thrilling mototaxi ride from the port to our hotel. After leaving our kayaks in the care of the Coast Guard, we commandeered six of the rickety machines to haul the team and gear through the streets, packed with pedestrians, neon lights, strange wonderful smells and sounds of life.

Under the disapproving and confused stares of the hotel staff, we piled our odiferous gear at the front desk in the marble-tiled lobby. Erich quickly arranged for a storage room so we wouldn't have to haul all the gear upstairs to our rooms. Air conditioning, running water, sheets, pillows and solid walls were quite a step up in the world from our digs in Galilea. Several large in-your-face formal signs along the walls of the lobby and stairwell notified patrons in English, Spanish and stick-figures that having sex with children is against the law and not welcome here. I envisioned a disappointed pedophile walking through the lobby with his victim, seeing the signs and slapping his forehead in surprise, then turning to leave.

Miguel stayed with his brother, a well-regarded magistrate in Pucallpa, but still managed to make it to almost every team meal during our two-night stay. He even showed up to the hotel each morning for our complimentary breakfasts. His last superfluous task with the expedition was to ride in the mayor's speedboat when Rambo drove it back upriver to Atalaya, but he seemed in no hurry to get started.

Our layover day in Pucallpa was filled with shopping (tuna, crackers, mustard, hard rolls) and cleaning and organizing gear. Erich met with local media for interviews, organized his photos and corresponded with National Geographic magazine. Jeff and Ian measured and packed baggies of the protein powder we relied upon for the calories that kept our bodies running. I washed all three tents and ground covers at the hotel pool, draping them on the fence and bushes to dry. Jason hired a woman to wash all of our clothes and Pete did some kayak maintenance.

Che Guevara and Alberto arrived in Pucallpa by truck on May 24, 1952, having spent eight days traveling from Lima in the company of two brothers they nicknamed the Cambas, suffering a broken axle and other misadventures along the way. The road was closed in several places due to heavy rains, and though the travelers did their best to talk their way through the barricades, when they reached a small town called Nescuilla on the evening on May 22 they could go no farther. As usual though, Che and Alberto found a workaround.

> *"The road was still closed the next day so we went to the army post to get some food. We left in the afternoon, taking with us a wounded soldier who would get us through the army roadblocks. The strategy worked: a few kilometers down the road, when the rest of the trucks were being stopped, ours was allowed through to Pucallpa, where we arrived after nightfall. The younger Camba paid for our meal and to say goodbye we drank four bottles of wine that made him sentimental and promise us his eternal love. He then paid for a hotel room for us to sleep in."*

Per our team meeting in Galilea, I'd told the others that I would clear the air with John in Pucallpa, so at 6 a.m. I taped a note on his hotel room door asking him to find me at the pool so we could talk. I spent most of the day around the pool cleaning all of our tents and gear with a good supply of salami, crackers, rum and cola. John was nowhere to be found. It was mid-afternoon when I spotted him coming into the lobby from the street, headed for his room. I caught up and asked him to join me poolside for our chat. The rum had mellowed me a bit, ushering a more passive, and hopefully therapeutic perspective for our talk. I remembered the approach David and I agreed upon back in the Andes and dusted off whatever problem-solving techniques I learned working two decades in the mental health field. This was going to be one big mushy Leo Buscaglia hugfest. The atmosphere was going to be a giant pillow of acceptance to catch and hold John's anger and frustration and smother it in cuddly side-hugging manly I'm-okay-you're-okay warmth. My whole essence was a plethora of bring-it-and-I'll-take-it, of which the rum and coke was an essential element.

I asked about John's level of satisfaction with the expedition as a whole and about me and my actions specifically. I emphasized how much I valued his perspective and urged him to speak freely.

"Some people aren't interested in hearing about your views on religion and politics," he said.

During the last few hundred miles, we had several long, academic discussions about religion and politics, all very civil and interesting. John had kept quiet during all of them.

Fair enough. He made a really good point. No more religion or politics.

Next, he complained about not having a GPS unit. Before leaving Texas, I purchased an expensive handheld GPS, which I kept attached to my deck, and bought a second unit for a backup, which Jeff carried. Given John's recalcitrance in operating a simple walkie-talkie, it didn't seem like trusting John with a $500 GPS was a good idea. Nevertheless, in the spirit of teamwork, I promised to purchase another GPS at the very next GPS retailer we came across as we kayaked through the jungle.

John then said he was particularly annoyed with Jeff and Ian's habit of doing sprints now and then. They'd sometimes sprint for a hundred yards or so just to wake up from the doldrums, and then drift until the rest of us caught up. For some reason this really got under John's skin. However, if the sprints were getting in the way of team cohesion, then I'd ask Jeff and Ian to stop. Done.

After he exhausted his list, I urged him to tell me what he thinks I could do to improve things on the expedition. He said he couldn't think of a thing. My turn.

"So, what was all the swerving about when we were coming into Pucallpa last night?"

John explained that he was tired of the boys sprinting ahead of him. I told him that they were just trying to ride our wake and stay close in the dark, leaving alone the question as to why Jeff and Ian's sprinting bothered him so much. He strongly disagreed. He was sure Jeff and Ian had been determined to sprint ahead. I ventured that, without asking them, John couldn't be sure of Jeff and Ian's intentions. And anyway, I simply couldn't tolerate teammates cutting off or ramming one another.

I told John that there was enough to deal with out here in the middle of the desolate Amazon without contention between team members, and I expected everyone to voice a grievance before assuming the worst of one another. John agreed, but only after insisting that I talk to Jeff and Ian about all of this, too. Of course.

Halfway through our talk, Jason approached with a bottle of Johnnie Walker to share. His signature smile faded as I asked for a few more minutes alone with John.

"No sweat. I'll check back later," Jason said, turning to walk away.

"Hold on!" I called out. "Leave the bottle." After all, the rum was gone.

From here on out, I told John, I'd be confronting him immediately if I saw something like the boat ramming, or if he and Pete paddled away from the group. To that end, we agreed to a new rule for the expedition: no one would ever be more than a hundred yards from the rest of the group at any time.

"You going to tell this to Jeff and Ian, too?" John asked again, in a rather accusing tone. While pouring the whiskey, which John declined, I agreed to explain the new policies to everyone at dinner that night and gave him a final opportunity to raise any more concerns.

Finally, just as we had things wrapped up, amidst the warm glow of fermented grains, I ventured to suggest that John take the bow position in their kayak, because he was a stronger paddler than Pete. John took the bait and the problem of their kayak paddling out of synch was solved. I promised to break the news to Pete, so John wouldn't have to be the bearer.

Over steak dinners, I announced with a great flourish of authority the new 100-yard rule and that our pace would be dictated by the slowest boat, which, I pointed out, was usually mine by the end of the day. We would all take the same route around islands and if anyone thought one way was better than the other he was to speak up. I also expected people to air their grievances as soon as they became apparent and clarified that Erich and his team were here to chronicle our expedition, not to support it. The help they gave us hauling gear, buying bottled water and arranging safety was nice, but not to be expected or taken for granted. I checked for any further concerns, ideas or feedback and got none. Heads nodded around the table and I felt a weight had been lifted.

That night, the mayor invited us as guests of honor to the Miss Pucallpa Beauty Pageant. Dancers opened the extravagant show, held outside on the Plaza de Armas. The square was packed with thousands of costumed revelers re-creating the history of this part of the world. They played out the legends of hunters and gods, of conquistadors and Catholic missionaries. Throughout the impressive show, romance, passion and a celebration of jungle life was the foremost theme. Lots of steamy jungle passion. Barely dressed dancers flitted and jumped around stage, artfully impersonating hunting and romance. Eventually, some diminutive women in traditional fiesta dresses pulled us up to dance in front of the crowd. If they were expecting any resistance, they were sorely let down. All of us jumped up and did our best to adapt our Texas swing and two-step styles to the Amazonian beat. It was great to see even John smiling and twirling with the gals, and I hoped his burden had been lifted, as well.

Jason takes Pucallpa by storm. Photo by Erich Schlegel

Eventually, Jason took charge and led a huge samba line, complete with hip swings and kicks. We all had a great time, with the exception of Pete who stayed at the hotel.

"Becky just wouldn't understand," he said.

Worn out from the river and a busy day, I left the team to handle the pageant and shuffled back to our hotel. Walking down the hall, I noticed my note still taped to John's door, still folded and unread.

Queen of the Amazon. Photo by Erich Schlegel

Our final morning in town, Miguel showed up for breakfast on cue. The previous night, while out on the town with Erich's crew, he had repeated his vanishing act when the bar tab came due, and it was apparent to everyone but Miguel himself that he had worn out his welcome. After breakfast, Erich and César went in search of a boat and motor to purchase. Miguel towed along.

Just the day before, Rambo hit up Erich for $3,000 to return the mayor's boat to Atalaya. Erich figured the fuel would be a few hundred bucks, at most, but he got caught in the Amazon Shuffle and didn't have a leg to stand on. He didn't know the cost of gas along the route, nor did he know the amount of fuel it would take to muscle the fastboat back upriver. Being a man of his word Erich protested, but eventually paid the money, even though it took a massive bite out of his budget.

Hey, Mister Tallyman... At the port of Pucallpa on the Ucayali River. Photo by Erich Schlegel

Erich was still wincing from that shakedown when Miguel, with impeccably poor timing, came to him with his hand out, asking for $350 for a plane ride back to Atalaya. When Erich asked him why the hell he hadn't caught a ride back to Atalaya with Rambo the day before, Miguel merely shrugged, complete with palms facing forward and eyebrows raised – a full 10 on the shrug scale. Erich sent him off empty-handed, with a few choice words seared into his backside.

After two days in the lap of luxury, eating at ice cream parlors and steak houses, we were well rested and antsy to get moving. As the river grew in size, so did the boats. Bigger tugs pushed bigger barges. Tightly packed three-story wooden riverboats ferried crowds up and down, stopping at ports of call. More pecky peckys rattled and thumped here and there, with the occasional curious gawker daring to come in for a closer look. For the well-heeled, fastboats with large outboards screamed up and down the Ucayali. Our team wasn't exactly pulling over for group hugs, but for the first time everyone seemed to be getting along, and all seemed well.

Escorted out of Pucallpa by the Coast Guard. Photo by Erich Schlegel

The Honeymoon is Officially Over

"Some people create their own storms, then get upset when it rains."
Unknown

Each morning and evening we pumped water through the filter, though Pete insisted on using purification tablets exclusively. He dropped a couple of pills into a plastic bottle of gritty river water, waited a few minutes then bottoms up. Pete didn't trust the charcoal and porcelain filters the rest of the team relied upon. With a background in chemistry and a job with the water department, Pete felt it best to zap any potential bugs with high-test chemicals. I took a different view. Raw sewage spilled out of every town on the river, and I figured a chunk of feces was still a chunk of feces, regardless of whether the bacteria it harbored were dead or alive. For the rest of the team, it seemed best to simply strain out the poop particles and dead fish parts before drinking.

From Pucallpa, it took us a few really long days to reach Contamana. While Erich and crew appeared now and again, we adopted an independent perspective and maintained an easy pace, as the humidity and mosquitoes steadily gained prominence. Though kept at bay by heat and pounding sunshine during the day, at some point in the evening the supper bell clanged and the blood-suckers came out in clouds. We learned to be inside our tents before dusk.

We also learned to drag our kayaks well away from the water's edge and stake them down against waves generated by midnight barges. Sand and grit were constant irritants. Dewy mornings left us with gritty wet ground covers so we got in the habit of rinsing the sand off the nylon covers then packing them on deck to lay out and dry during our mid-morning break. Little adaptations went a long way.

During this 160-mile section from Pucallpa to Contamana the fatigue of the expedition caught up with me like an anvil in a punchbowl. The cumulative effect of the past two months left me worn out, though I ate twice as much as everyone else. I was the only team member who had paddled the 512 miles of whitewater on the Mantaro, plus another 650 miles so far on the Ene, Tambo and Ucayali. These miles piled up against my depleting fat reserves, worn muscles and emotional resolve each day. Stopping was never an option, but the going got tougher with each stroke.

I began joining Jeff in his mid-day coffee breaks, but even with the caffeine jolt and ample calories my energy and speed would wane around 3 p.m. One day, as I faded back a couple of hundred yards behind the tandems, Jeff and Ian drifted until I was able to catch up, and then matched my slow pace. Up ahead about half a mile, well beyond our new 100-yard proximity rule, Pete occasionally looked back and stopped paddling, but John just kept hammering away in the bow. Eventually, we were able to catch up to them. I thanked them for holding up and, pulling up alongside them, told John that I was the weak link in the team and would really appreciate it if they stayed close.

"Does that go for everyone?" he shot back.

I tried to maintain the kumbaya tone I thought we'd established in Pucallpa.

"Yes, in fact Jeff and Ian stopped to wait for me just now as y'all continued downriver."

"Well, they did *today,*" he grumbled. Sticking to my promise to call out grievances and address them as they arose, I asked John what he meant.

"They didn't stop before!"

In a steady tone, I reminded John that we all agreed to stay within 100 yards of one another at the speed of the slowest boat, and that we were alone in the middle of the Amazon with no one to rely upon but one another. Pete, apparently weary of the therapy session, brusquely interjected that we were all going to start working as a team and it wouldn't happen again. Case closed.

Freshwater dolphins became a familiar sight, both the little blue ones and their larger pink cousins. We learned to look for them at a twain or in eddies below islands. Small pods of the little blue guys shot 10 feet out of the water in fanciful arches. The pint-size blues varied in size from four to six feet long and closely resembled their saltwater cousins. Not at all shy, they often shadowed our kayaks with a childlike curiosity. The larger pink dolphins were about the size of your average heifer and just as active, keeping their movements to lumbering partial breaches and impressive whale-like blow-holing. From the shore at night we often heard the pinks checking us out, blowing air under the moonlight. Their snouts had a bumpy protuberance between the tip and the forehead, which seemed to extend to the dorsal fin, giving them a prehistoric look. Dolphin sightings really perked up my spirits, breaking up the somnambulant doldrums of ultra-distance paddling.

During one of our daily calls, Barbara reported that she hit a brick wall after approaching National Geographic for additional funding. The delays on the Mantaro and frequent drip-drip of extortion drained our coffers. National Geographic contributed $35,000 to the expedition, but they didn't simply hand over the dough. Several strings were attached, including a complete blackout of all news regarding the new source and no contact with any other media without permission from National Geographic. National Geographic claimed that they wanted to reveal the new source on their own terms, within their own timeline.

Consequently, we gave up a huge amount of press coverage, not to mention other potential sponsors that may have jumped at the chance to support the first full descent of the world's largest river from its true and newly discovered source. According to Piotr, National Geographic nixed a planned joint press conference with Rocky and our team in Lima before the start of our expedition, which potentially would've provided us with a center-ring performance for the world to witness. Despite its brand as the world's leading popular geography organization, National Geographic preferred *Area* to come out with the news first—and thus assume the risk that he might be wrong—before publishing the news in their flagship magazine. The problem, of course, was that Rocky was busily running rivers all over the Andes and hadn't the time to complete his academic paper on the new source, much less shepherd it through the peer review process. Justifiably, being the first to run the five major source rivers of the Amazon was his primary goal and the issue of the most-distant source simply popped up at the last minute. Rocky, too, was under a gag order. He signed away the right to talk about his own discovery in exchange for a National Geographic grant.

We were in limbo, forbidden by our contract with National Geographic from so much as mentioning the words "new" and "source" in the same paragraph, and certainly not to news organizations or potential sponsors. Our team had gone to great lengths to keep the Mantaro source a secret, even as we posted daily trip reports online which, ironically, was another requirement in our contract with National Geographic. Given the substantial financial sacrifice we'd made to uphold our end of the bargain, I figured National Geographic could let loose of another $10,000 to insure the success of the expedition and, incidentally, their big source scoop.

National Geographic told Barbara that they didn't know anything about a proposed press conference in Lima and openly doubted we had spent as much as we claimed. Our grant was based upon matching funds, which we more than doubled. Perhaps Piotr simply proposed the press conference without first clearing it with his friends at National Geographic and later called it off. Either way, National Geographic had no immediate plans to promote the expedition, and we were contractually prohibited from breathing a word about it to any other media.

Geographic did, however, ask for an itemized list of expenditures, with receipts, to prove we had actually spent what we claimed. Fortunately, I was quite the taskmaster when it came to saving receipts, preserving each and every one of them, from dinner tabs to mototaxi rides costing only a few *soles*. I even logged the cash payments for shakedowns large and small. Unfortunately, I was in the middle of the Amazon jungle with a good number of those receipts. Another large collection was in a storage room in a hotel in Lima, and yet another stack was back in Texas. I was in no position to conduct this audit and get the requested paperwork to

Washington D.C. Barbara eventually convinced National Geographic to accept a detailed narrative instead, which she and I hashed out over the satellite phone at four dollars per minute. After his own call to Washington a few days later, Erich reported that the magazine was no longer sending Joe Kane for the interview, but would send someone else, at some undefined point in the expedition.

Chapter Twenty-One
Contamana

Day 58, Mile 1,243

"My asthma showed no sign of abating and I had to take the drastic measure of getting asthma medicine by the banal method of paying for it. It relieved the attack slightly. We looked out dreamily beyond the river edge to the jungle, its inviting greenery so mysterious. My asthma and mosquitoes restrained me somewhat, but virgin forests were so compelling for spirits like ours that physical impediments and all the nascent forces of nature only served to stimulate my desire."

Che Guevara, May 1952, Ucayali River

While the jungle provided a somewhat monotonous green billboard, it was a billboard I could watch forever. The unbroken lushness inevitably pulled us into deeper thought once we were drained of the energy required to talk. Erich's team was several days behind us, finding a boat to purchase in Pucallpa and gearing up for the journey, leaving us untethered through the quiet miles, thick with water infused air and breezeless heat.

Contamana rose up at the base of some uncharacteristically tall hills on river right, the first geologic formations of note since Atalaya. The lofty knolls looked like something right out of a King Kong movie, draped in greenery and dripping with moisture. It was the perfect terrarium: always moist, always hot, always sunny, always beautiful. While the immense flat jungle was never tiresome to behold, it was a nice variation to see the tall emerald hills rise against the infinite indigo sky.

The street lamps along the Contamana promenade glared in contrast to the oft-changing mud and dirt banks synonymous with the tiny villages we passed. Given a few short years of neglect, the patient advance of the jungle would subsume the little city, as it had overrun the civilizations that came before. Constant vigilance is required to keep at bay such a powerful expression of nature.

The gate to Contamana. Photo by Erich Schlegel

As the sun set a speedboat bristling with small arms roared up. The driver cut the twin outboards and the man clearly in charge, Ensign Ray Paredes of the Peruvian Coast Guard, addressed us in English, "Are you the National Geographic expedition?" We exchanged pleasantries with Paredes and his colleague Ensign Roberto Yokota and followed the Coasties to a massive military barge moored on the city waterfront.

At the barge, 30 young troops jumped at Paredes' command, hauling our kayaks aboard. We were shown the shower facilities and instructed to pitch our tents on the deck. Once cleaned up, with gear stowed and photos taken with all the recruits, Paredes took us into town for fried chicken.

The plaza was just setting up for a festival celebrating the 100th anniversary of the town's incorporation. The city of about 15,000 was clean and well maintained, with concrete streets, curbs and an inordinate number of beautiful women. Ensign Yokota, who went by "Toshi," met us at the café, where Paredes introduced us to a tall young man named Yonel Guzmán. He spoke excellent English with an exotic flair, like a young Ricardo Montalban. Yonel was the sort of guy who could weaken a woman's knees with a simple "allo." A real knicker-dropper.

Yonel belonged to one of the most successful families in Peru and was in Contamana on business, managing his family's timber interests up and down the Ucayali. Though only in his late twenties, he was the epitome of a perfect continental gentleman of the old world, with impeccable manners and humor. He was immensely knowledgeable about environmental concerns in the Amazon and spoke at length about his policy of preserving stands of old growth timber, even though the Peruvian government granted his company rights to harvest them.

When Yonel was 21, his father sent him to New York City to learn English. After a couple of years bouncing between New York and Houston (he spoke pretty fair Texan, as well as English) Yonel not only mastered the language, but established a network for importing milled lumber for flooring. Yonel's strategy was to leave the old growth standing and build a market for fast-growing species such as lupuna, ipe and a variety of cedar known locally as *pollo sangre*—chicken blood—for its rich red color. The timber is harvested, hauled through the jungle to his mill in the town of Orellana, and then carved into large chunks to be shipped by barge to Pucallpa where a second mill turns the planks into finished tongue-and-groove flooring. Yonel had his hand in every step of the process, to the point that he knew exactly how many pieces of four-foot flooring could fit in shipping containers bound for China, Africa, Europe and North America. His biggest obstacle was getting the cut logs through the jungle to the Ucayali River.

In three days, Yonel was scheduled to meet with the Peruvian President about establishing a railroad through the forest. Then he and his brother would travel to Russia to check out a used locomotive for sale. The train engine would make the journey by ship from Russia, across the Atlantic Ocean and 3,000 miles up the Amazon to Yonel's mill. We accepted his invitation to tour the operation down river in Orellana, then picked up ice cream on our walk back to the Navy barge for the night.

Through the Land of the Isconahuas

Orellana – Day 59, Mile 1306

Though Orellana wasn't far as the crow flies, a series of 20-mile hairpin curves made it a long days' journey. The GPS mounted on my deck was a blessing and a curse, noting the river that stretched out miles in front of us, before doubling back on itself to land a mere 100 yards from where we currently paddled, though in the opposite direction. The Amazon was truly a living being, constantly growing and changing, cutting a shifting path through the jungle, which yielded to the river's will without ever giving in.

In spite of all our efforts to improve team cohesion, John seemed to be in a perpetual state of irritation as his and Pete's kayak kept their distance while we paddled. The rest of us were relishing this amazing adventure, but John just seemed bound and determined to be unhappy. We were dumbfounded, but I had no interest in adding John's perpetual victim status to my to-do list to fix.

Just north of Contamana the hills of the Sierra del Divisor rose abruptly out of the flat carpet of jungle between Peru and Brazil. Declared a national park in 2015, the 5,229-square-mile reserve is larger than Yellowstone and Yosemite combined, and protects the only mountainous region in the lower Amazon jungle. The new park is the final link in the Andes-Amazon Conservation Corridor, extending for more than 1,100 miles from the Brazilian Amazon to the peaks of the Peruvian Andes.

The protected area on the river's right bank includes the ancestral territory of the Isconahua people. The lands on river left, from the dense jungle west to the piedmont of the Andes, are home to the Sencis, who read stories in the stars just as other cultures have for millennia. I was able to locate the Vaca Marina/Manatee (Scorpio), Pijarre/Arrow (Regulus) and Capaygui/Alligator (Sirius)—constellations that Lieutenants William Smyth and Frederick Lowe recorded when they explored this river in 1834 in the service of the English crown. It's comforting to know that people through the centuries have stared up at the same sky and seen familiar shapes in the stars we all share.

This isolated *cordillerita*, or small mountain range, rises from the jungle floor to a height of 2,300 feet and harbors plants and animals adapted to the unique environment. Eighteen species of monkeys are found within the Sierra del Divisor, and the region is the only known habitat of the red-faced "Red Huapo" monkey. Over 1,000 new species of plants have been discovered in the sierra. Though protected, the park lacks the staff needed to patrol the vast area, and illegal logging is endemic in the park and its Brazilian counterpart, the 3,268-square-mile Parque Nacional da Serra do Divisor.

Because of our goal to reach Orellana by day's end, we didn't stop for lunch, but drifted while we ate. Just before dusk, several playful dolphins joined us, jumping in the flat water against the setting sun. The GPS didn't show Orellana so we were flying by the seat of our pants until the lights of the town appeared around a long right-hand bend, several miles downriver. Ian found Yonel's large sawmill in the dark and a couple of us climbed the steep bank and were led to his office. A few minutes later, our sweaty host popped in to inform us that the *fútbol* game in which he was playing with his employees was almost over and he'd join us momentarily. We followed him to the lighted pitch, where a heated game was underway, with Yonel in the middle of the action.

Following the game, Yonel directed us to leave our kayaks at his mill for safety, and had his fastboat driver take us into town, where he'd already arranged our hotel rooms. We planned to meet up after we all had a chance to rinse off a bit for supper. My cave-like room with a caged window, concrete walls and floor was conveniently located next to the town's loudest disco, which was just finding its pace. Pete, not feeling well, opted out of our dinner with Yonel, and John was a no-show, as well. Compared to Contamana, Orellana had seen much better days. It was difficult to tell whether buildings were being torn down or rebuilt. The concrete streets were crumbling and huge craters appeared here and there with no apparent purpose other than to keep us alert. Even the mototaxi drivers were cautious.

At Yonel's favorite outdoor café, he asked, "What'll you have?" We were tired of chicken, potatoes, rice and plantains, so without hesitation, I ordered a hamburger or a steak. Jeff and Ian doubled that order. Yonel interpreted this for our waitress, who immediately shook her head. Yonel relayed the obvious to us.

Not wanting to waste any more time, Jeff asked, "Well, what *do* you have?"

"*Pollo con papas, arroz y platanos.*"

Yonel looked at our droopy faces and added, "But, you can have the chicken fried or grilled! It's really great here." When the waitress left with our orders, Yonel admitted that he really missed American pizza and hadn't managed to find a decent cheeseburger anywhere in Peru.

Over dinner we talked about our experiences to date, including being guests of honor at the beauty pageant. Turns out Yonel dated the utterly beautiful Miss Peru, who hosted the Miss Pucallpa pageant in a shiny silver-sequined mini dress shrink-wrapped around her more-than-acceptable figure. She had been the runner-up in the national pageant and inherited the title of "Miss Peru" after an unfortunate incident involving the crowned winner. While Yonel's command of English was quite excellent, his highly cultivated manners left him at a loss to convey what happened between the dethroned winner and a pageant judge, but from what we gathered it was wrought with details befitting a Trump-level sexual encounter.

Sea breve, por favor
(Please, keep it brief)

"A hangover is just your body reminding you that you're an idiot."

It goes against the laws of nature and morality to wake up hungover after a night of sobriety. It's just plain wrong. This morning was payback for all the nights spent drinking, after which I woke up feeling great. I'd spent a fevered night on sweat-soaked sandpaper sheets beneath a phlegmatic fan pretending to push stale air in my general direction and failing impressively to provide even minimal relief. My earplugs proved helpless against the onslaught of noise emanating from the disco on the other side of my wall. Donald Rumsfeld himself couldn't have devised a more perfect torture. I would've turned over state secrets for a moment's relief. Pete reported the same symptoms, which seemed indicative of dehydration.

Yonel graciously covered the cost of our rooms and his boatman motored us back upriver to the sawmill where we spent a few minutes packing our kayaks on the muddy bank near piles of extraordinarily massive logs. The painful jungle-rot, faux-hangover and grimy-gut left me hobbling around like sailor stumbling back, broke and broken, from a highly successful shore leave.

The sawmill was an OSHA nightmare and a Libertarian's wet dream. Under the tin roof of the mill house, workers in sandals or bare feet and soccer shorts moved gargantuan logs into place with front-end loaders. Once a log was on the cutting surface, two young guys scrambled around to clamp it down before firing up the band saw, a circular ribbon from hell about six inches wide and 16 feet in diameter. It was powered by a diesel engine and thrown into action with some sort of clutch mechanism that made an ungodly whining roar whenever it was literally kicked into place by one of the bare feet. Once the sheet metal blade was whirring, a worker kicked another clutch to start the band moving down the length of the stationary log. The speed of the blade down the log was controlled by another lever held in place by one of the barefoot boys as he walked along the length of the log. If that thin blade were to break while in action and go flying, surely some body part would instantly be separated from its owner.

Once the entire log had one flat side, the prehistoric blade and machine were swung out of the way. Another young man handed a cable underneath the newly sawn log. At the end of the cable was an obscenely large hook from a slasher movie.

The saw operator hammered the hook deep into the newly cut timber. When the cable was pulled by another machine, the big log rotated and landed on the flat side with a massive *whomp!* The young guys remained nonchalantly perched within inches of the multi-ton log as it flopped over like a breaching whale. They seemed utterly confident in their safety, as if they'd never given a thought to being crushed or dismembered, though I'd wager Orellana is full of their one-handed, one-footed former employees. With the log lying flat, the process was repeated three more times until what remained was a perfectly squared-off chunk of wood with enough volume to frame a two-story house. All I could think of was what emergency medical care was on hand to treat a crushed limb or severed body part. The people of Peru are truly amazing in their skills, intelligence and innovation when it comes to making things happen in the most remote regions.

Jeff, Ian and I found Yonel in his tiny office building to thank him for his generous hospitality. Behind him hung a framed slogan, *Sea breve, por favor* ("Keep it brief, please"). Upon returning to shore we found Pete and John had already departed. We caught them drifting by downtown Orellana two miles downstream. From across the river, Jeff called out for them to join us in the faster moving current, but got no response. Pete finally waved us off, so we continued to drift and wait for them.

Eventually, they started paddling and joined us five miles later when we stopped on a gravel bar for breakfast. Before we all finished eating, John and Pete started loading up to head out. In the easiest-going voice I could muster, I told them that from here on out, I wanted all team members to depart at the same time. They sat in their kayak while the rest of the team packed the cooking gear. Everyone ignored John's sarcastic "Are you ready, yet?" as we loaded up and launched.

After lunch they left before us, yet again, and we needed almost two miles to catch up. By three in the afternoon I was lagging well behind the two tandem boats. My energy level was at rock bottom, though my spirits were high. I simply didn't have enough fuel reserves, having dropped a lot of weight. Around 5 p.m. I was lagging far enough behind that Jeff and Ian stopped paddling for me to catch up, once again. I just didn't have it in me to go faster. They told Pete and John, paddling alongside them, that they were stopping to wait for me. When they didn't show any sign of letting up, Ian repeated his statement to make sure he was heard. Still no response. John, in the bow, didn't so much as turn his head. Jeff and Ian stopped paddling, and by the time I struggled up to them Pete and John were almost a mile ahead. I kept a close eye to see if they would look back for us, but they just kept hammering along at a strong pace. Jeff and Ian putzed along at my snail's pace.

I was spent and pissed off. Here we were, all alone in the Amazon with no chance of rescue if we were to encounter trouble, and these two guys were behaving like spoiled children. They couldn't even bother to look behind to see if the only people on the planet that could come to their aid were within sight. On they paddled as the three of us watched.

I did my best to ride Jeff and Ian's draft, but there was no way we could catch them. It was after five o'clock, the usual time for us to find a camp spot. After some discussion and attempts to raise Pete and John on the walkie-talkie, I made the call to pull over to a really nice-looking beach on river left. We kept an eye on their kayak through the binoculars. Not once did any of us see Pete or John look behind. Who knows what was going on in their minds, but one thing was for sure: they didn't care to be around the rest of us.

The three of us had no earthly idea why they were brazenly flouting the agreement we'd made in Pucallpa or what the hell John was mad about. I searched my recent memory for anything I may have done to anger them, but came up with nothing. In fact, I had gone out of my way to include John in all discussions and ask his opinions. I notified Barbara of the situation and she reported that their satellite transponder showed them stopped on the right bank about three miles downriver. She never heard from them that night. Records later indicated that they each used the sat phone to call their respective loved ones back home that evening.

Ian, ever the peacemaker, wasn't all that thrilled with us stopping up river of Pete and John, though he understood my reasoning. Jeff was all for it. He had long ago grown tired of John's recalcitrance and his respect for Pete was diminishing. True, we could have continued downriver into the evening with hopes of catching them, but I was worn down. It felt like I was dealing with a couple of teenage girls "ghosting" me, for who knows what.

After a stormy night I gave Barbara a rare morning call to see if she'd heard from Pete and John. As a matter of fact, she'd just gotten off the phone with Pete. She gave him a piece of her mind for splitting the team and not calling her with their status, as we agreed. Once we paddled down to their spit of land, I calmly and frankly told them that this simply could not happen again.

"It won't," Pete quietly replied.

John didn't respond.

They didn't have a stove or water filter, therefore dined on cold snacks and sipped Pete's treated river sludge. Several times they scrambled to move up their diminishing sandbar during the night as the river rose with the storm. Neither of them appeared to be in good humor, especially after Barbara chewed off a good chunk of Pete's ear. We all stuck together the rest of the day, though the mood was thick.

The Cat is Out of the Bag

"Three may keep a secret, if two of them are dead."
                                                    Benjamin Franklin

Just when we thought there was enough drama, Barbara received a flurry of calls and emails from Grayson Schaffer, an editor at *Outside* magazine who caught wind of the new source and Rocky's dismissal from the expedition and wanted to interview me.

What the hell?

Our respective contracts with National Geographic specifically forbade any communication with *Outside* or *Discover* in any way, shape or form. Rocky had yet to submit his completed study of the new source to *Area*, but launched several email missives to National Geographic, accusing the organization of misleading him and stating, among other things that, "I WILL COMPLETE MY DESCENT BEFORE WEST… and therefore have claim to the FIRST DESCENT OF THE AMAZON."

Rocky was upset and wanted vindication.

Schaffer asked David Kelly point-blank about the new source and other details about our expedition. David, now the official voice of the expedition, politely replied that we were under contract with National Geographic and were not at liberty to discuss anything with him. Their interactions were cordial, though Schaffer was persistent, baiting David to respond to Rocky's allegations of my backstabbing, manipulation, deception and conspiracy against him. To his credit, David never took the bait, though he requested permission from Rebecca Martin, National Geographic's director of explorer programs, to participate in an *Outside* interview to defend me and the expedition against Rocky's accusations. Rebecca declined his request, so David held the company line.

Schaffer then made the ballsy move to call National Geographic communications director Barbara Moffett for confirmation. During their brief conversation, Schaffer alluded to the fact that he and David had been talking about the new source of the Amazon, without bothering to mention that it had been Schaffer, exclusively, doing all the talking. Moffett responded that there had been a claim of a new source, for which National Geographic was sponsoring an expedition, but the new source had yet to be verified. That gave Schaffer the confirmation he needed. Moffet then confronted David, who explained that no one on our team had breathed a word about the new source.

The irony wasn't lost on me; the organization that forbade us to discuss the expedition with any media was the same that released the information to its arch rival—and then continued to restrict our access to any other media. This was the last time anyone from National Geographic contacted the team during the expedition, save for Erich, who was in contact with the photo editor of the magazine. Barbara continued to send updates to National Geographic, but the organization never acknowledged them or responded. With thousands of miles still yet to kayak, we became the awkward one-night stand who expected something more than "catch and kill" journalism in the light of day.

As the days wore on, the politics of our expedition wore me down, more than the Amazon could ever do.

Man's gravitation towards the familiar, with its guaranteed safety ensconced well within the folds of the predictable, leads him to a reluctance to veer far from home, even if home isn't hospitable. The devil you know is far preferable to the devil you don't know. Gradually, while still missing the well-known prairies and hills of Texas, our new familiar was becoming the daily campsite and banks of the Ucayali. I felt comfortable and I suspect some of the others experienced the same. The Mantaro was far too violent and dangerous to settle into any level of consolation, but the Ucayali lulled me into a relaxed existence.

Perhaps this is what it feels like to have an affair, feeling comfort in the arms of two, though this mistress wasn't to be trusted.

Chapter Twenty-Two
Nothing but The Best

Day 61, Mile 1,418

"There are a lot of mysterious things about boats, such as why anyone would get on one voluntarily."

P.J. O'Roarke

It had been almost a week since we left Erich, Jason and César in Pucallpa. Barbara relayed messages from them that we should slow down to give them a chance to catch up, but none of us were interested in lying around. Eventually, they showed up in their colorful pecky pecky just before our morning launch, two days below Orellana. The bright wooden boat sported a flat tin roof with "The Best," in English, painted on its leading edge. Why keep looking for a boat once you've found the best? The superlative little craft was pushed along by an unmuffled Honda outboard steered by a six-foot-long piece of rebar with a propeller attached to the tip. Erich paid about $3,000 to purchase the boat, which he loaded with food, water, soda, rum and other much-needed supplies. It was really great to see friendly faces.

The YETI cooler Erich dragged all the way from Texas, much to my chagrin, turned out to be the most popular piece of luggage, due to its ability to keep ice frozen for several days on end, meaning we had cold rum/cola cocktails to look forward to around each day's end.

Jason immediately solved the mystery of my feet, which had become so painful it was all but impossible for me to stand: jungle rot. It was the first time we put a name to the ailment, which waylaid so many soldiers in Vietnam. His prescription, consisting of huge amounts of foot powder, sounded overly simple for such an acute problem, but he explained that my feet needed to dry out each day in order to heal. Jeff had an ailment of his own to report: a hard cyst about the size of a walnut on his rear end. He cut a hole in his seat pad to accommodate it and kept going, with plans to have it treated in Iquitos two weeks downstream.

Tranquility at days end. Photo by Erich Schlegel

While Jason played doctor, César entertained us with bird calls and animal imitations. He could name any bird by its song, then cup his hands over his mouth and imitate it perfectly. Of special note was the ubiquitous *el alma perdida* – the lost soul.

Father John Augustine Zahm was a priest, scientist and explorer who inspired Teddy Roosevelt to launch his infamous River of Doubt expedition before getting dismissed for being a whiny pretentious ass. The two became friends and Roosevelt penned the forward to Zahm's book *Along the Andes and Down the Amazon*, which the intrepid priest wrote under nom de plume H.J. Mozans. In the book he wrote of this bird and the legend of its name.

> "It is related that a young Indian mother left her child in charge of her husband, while she went into the forest to collect balsam. Alarmed at her long absence, the man went in search of his wife, leaving his child behind.

When they returned to the spot where the child had been left, it was gone, and to their repeated calls, as they wandered through the woods to search of it, they could get no response, except the mournful notes of this little bird, which, to their over-wrought imagination sounded like papá, mamá — *by which name it is still known among the Quichuas of the montaña."*

Commodore César at the helm of The Best. Photo by Erich Schlegel

The propeller shaft on *The Best* was rattling inside its pipe sleeve like no one's business, so Erich and his crew planned to pull over for repairs in Juancito, a small town just before the Ucayali split into three main channels and a spaghetti soup of smaller ones. Most of the water and shipping traffic went left, into the Puinahua Canal, a wider and deeper waterway that, as any local will tell you, is nonetheless not the Ucayali proper. Naturally we planned to go right, taking care not to confuse the real Ucayali with Door Number 3, the Río Yanayacu. This was low-stakes navigation, as everything comes back together about 150 miles downstream.

Just before the split we rafted up with the *The Best* and a boatload of beer-swilling young gents in Polo shirts. They looked like a frat party on spring break, if frat boys carried 9-mm pistols in nylon holsters. Seems they'd heard about us on the radio in Pucallpa and decided to catch a puddle jumper to meet us somewhere on the Ucayali for a beer or 20.

As we kayakers took our leave and began working our way into the narrowing Ucayali, the party boys led the crew of *The Best* into Juancito, where a handful of barefoot mechanics hoisted the outboard up the steep bank to their shack and rebuilt the entire drive shaft with pliers, a machete and a piece of lumber. In the meantime, the boys were treated to copious amounts of beer, food and recreational gunfire in exchange for stories of their adventures. Their hosts wouldn't accept a dime of payment and refused to let them leave until all the beer and food was exhausted. On long expeditions, such as this, sacrifices have to be made.

Well stocked with beer and guns, Jason finds his homies on the Ucayali. Photo by Erich Schlegel

After 10 miles in the narrowing river I began to worry that we'd strayed into the Yanayacu instead of the Ucayali. We were moving pretty quickly through the channel, which was a scant 100 yards across in some places, but motorized traffic had all but ceased. Jeff, Ian and I pulled over to talk to a local about the river and confirm our location. I yelled out to John and Pete that we were stopping to ask directions, but they kept paddling. Just that morning they had gotten six miles ahead after again leaving camp before the rest of us, which resulted in yet another calm, but apparently fruitless, discussion about staying together. We approached the shore, where a farmer stood casually with a machete in each hand. I smiled and asked if this was truly the Ucayali. He assured me that even though the Puinahua

Canal was bigger, this was the true Ucayali River and not the Yanayacu several kilometers to the east.

This vital intelligence gathering was accomplished with a great deal of slow talking, repetition and gesticulation, all of which took about ten minutes. Meanwhile John and Pete paddled farther downriver. When we finally caught up I was at the end of my rope and let them have it with both barrels, throwing appropriate leadership techniques and all the kumbaya garbage out the door. I was fed up with their crap and let them know it, yelling and cussing that I'd been nothing but polite for the past two weeks but enough was enough. Out the window I tossed any semblance of diplomacy and group huggery. I simply wanted the team to stay within a safe proximity of one another as we made our way down the goddamned Amazon River! Hell, we agreed on 100 yards back in Pucallpa, just to satisfy John's pedantic definition. In defense, John said something about doing a "marathon pace" by keeping moving instead of constantly starting and stopping. When I asked about the relevance of pacing compared to the need to stay the hell together, he had no response. "Staying together supersedes anything else!" I hollered and reminded them, again, that I didn't need the goddamned stress of reminding them to stick together. They were behaving as if this was a safe, simple fishing trip in a local pond.

After that, things were quiet for a while. Really quiet. Dad had gone on a bender and everyone was on their best behavior until he came down again.

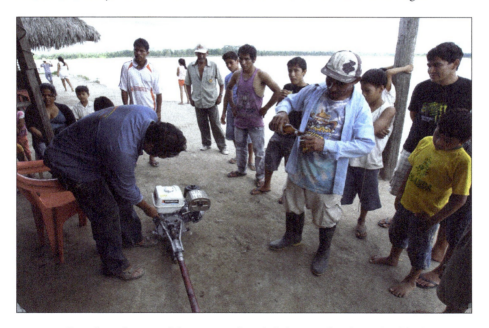

Friendly mechanics with beer come to the aid of The Best. Photo by Erich Schlegel

Camping next to a dugout canoe project. Photo by Erich Schlegel

La Gringa

Flor de Punga – Day 64, Mile 1588

*"And I can see you are an angel
With wings that won't unfold
Tune up your harp
Polish your old halo"*

Chuck Pyle

October 15. The 64th day of the expedition, and instead of celebrating my daughter's 14th birthday, I was slogging through the horse latitudes of the Ucayali River. It was one of the most depressing days of the entire expedition, a long, hot, weary grind. There was no treatment for my distress and I sure as hell didn't want to talk about it. We poked along at six miles per hour, a dismal pace compared to the speed we maintained upriver. John and Pete resumed their practice of keeping distant but I was in no mood to address it any longer. The path to their exit was well laid, and I simply didn't have the energy or will to care.

Every fish on the Amazon seemed to be ready for battle. Photo by Erich Schlegel

Phoning home in the evening from the trash-strewn muddy bank of Flor de Punga, I discussed normal suburban issues with my wife, then wished Isabella a happy birthday. She responded with teenage surliness, saying that she had a bad day but didn't want to talk about it. Really? I thought to myself, *You've* had a bad day with all that great food, rest, relaxation and air-conditioned comfort?

I'd had a tough day too, and my crappy, selfish mood spilled out through the airwaves, once again, to taint my daughter's festive night. Day after day, trudging through silent miles left us with nothing more than our own thoughts, and mine were back home. My emotions were in constant conflict between the wonderful experience of paddling the Amazon and the loneliness of being separated from my family. Was all of this worth months away from my girls—the irritation of dealing with surly partners, the pain in my torn shoulder, the jungle rot, sore muscles and the stagnant air? The imbalance left me irritable, so it was best to keep silent. I missed the daily routine of little nothings in my home that meant so much, now that I was away from morning coffees, bedtime stories, soccer games and micro-dramas. I missed the sweet, boring, wonderful, routine I took for granted. The call left me feeling terrible and missing my girls even more. Love is built and sustained in the mundane. Surrounded by the beauty of the Amazon and the excitement of our adventure, I couldn't help but miss the routine upon which our family's love had grown. Such is the plight of the explorer, longing for the road when at home and home when on the road.

Erich had a call as well, with the chief photo editor at National Geographic magazine, who told him the on-again/off-again meeting with Joe Kane was back on, this time in Manaus, about 1,000 miles downriver in Brazil. Afterward he told me the magazine may fund his coverage of the expedition all the way to the Atlantic, rather than pulling the rug out at Iquitos. This came as a surprise to me because Erich hadn't bothered to tell me he was planning on stopping before we got to Brazil.

Pete and I bunked in hammocks on *The Best* while César staked out a bed on top of the duffel bags strewn across the floor. The rest of the guys checked into the La Gringa *hospedaje*, which amounted to a cinder-block oven adorned with pornographic calendars. Erich bounced into town, talking with people, patting backs and finding the best local accommodations. By this point in the trip I loathed staying overnight in towns, what with the incessant noise and staring.

John liked the idea of getting a "good meal" in town, an indication of how precipitously his standards had fallen, but if he was happy, I was elated. For that matter, all of our standards had adjusted radically downward. We just didn't know how far.

While waiting for La Gringa and her staff to kill and grill the chickens, Jeff and Ian hit the bucket shower on the back porch while John made his way to the loo. It was a small, pitch black, one-seat affair set apart from the main building, with a single sewage pipe leading out to the street. Naturally, the pipe was clogged. The outhouse ceiling was just over five feet tall, requiring the patron to remain uncomfortably bent. Doing his best to squat while hovering close to the seatless rim in order to hit his target in the dark, John settled in and relaxed all the important parts. Just as all systems were "go," something from the putrid cauldron leapt up and bumped against John's delicate bum. Before he could rise, he got another light, albeit disturbing, bump. Launch aborted, John spun around and, careful to remain bent to avoid hitting his head, peered into the ghastly porcelain pit to find a frog wallowing in a nearly full bowl of human effluent.

Check please.

Perhaps the frog simply sought refuge from the possum-sized rat with whom Jeff shared the shower area. I chose the alleyway for my loo and a sudden thunderstorm for my shower. During the two-hour wait for dinner, Jason, Erich and I did some damage to a bottle of rum while admiring La Gringa's artwork. She derived the nickname from her fair complexion and the 12 years she'd spent in California. Her financial goal attained, she returned to her hometown to open her *hospedaje*. The pride and joy of her place was the mural of La Gringa herself adorning the wall of the large front room, which served as foyer, front desk, disco, bar and ad hoc café. The mural depicted La Gringa, about 45 pounds daintier, barely clad in a thong with ample, pale, gravity defying breasts. Depicted on the front of the thong was a pair of bright-red lips, puckered for a kiss.

After choking down supper, I made my way to *The Best* in the dark and thumped into my hammock. The boys polished off their anemic chicken tartar and a few warm soft drinks before heading to bed, while Erich and Jason stayed up chatting with La Gringa, who was a consummate hostess and very sweet.

Round about dark thirty, one of the town's sundry bone-thin dogs wandered into the *hospedaje* and came face-to-face with a monster-sized rat in the hallway, leading to a standoff for the ages with neither fleabag giving ground. Most likely, it was the rat with whom Jeff had graciously shared his bucket shower the evening before. The rat stood strong as the cur commenced to bark and bark and bark and bark and bark and bark and bark and bark and bark and bark and bark and bark and bark and bark and bark and bark and bark and bark. The scrawny dog wailed interminably against the unyielding rodent. Epic battles rarely transpire in a vacuum, and this was no exception. Every rooster in Flor de Punga voiced an opinion as bets were taken on the outcome.

Bunking down on The Best. Photo by Erich Schlegel

Throughout this hellish symphony, no one stirred from his bunk. Perhaps they thought a meteor would crush them into peaceful oblivion. Or maybe the planet would explode in a ball of flames. Stranger prayers have been answered.

Eventually, Jason stumbled out of his bed and stomped down the hall to kill the dog. As he tripped along, eyes bleary and feet bare, he called out above the din for all to hear, "I know every single one of you sons of bitches is awake and not doing a goddamned thing about this fuckin' dog!" The dog, perhaps familiar with imminent death in a man's eye, wisely ran off before Jason had a chance to properly murder it. The rat victoriously sidled up to the bar and ordered a dry vodka martini.

In the pre-dawn darkness we stumbled from our various lodgings down to the muddy bank alongside the fishermen of Flor De Punga as they headed out for their daily fish harvest. We were on the water just before the sun began its battle with the thick fog clouding our route. The midnight storms did nothing to cool the air, leaving us all damp, unrested and irritably quiet.

Pete, feeling bad during a break just upriver of the Maranon. Photo by West Hansen

We made nearly 60 miles that day, owing to our early start, and for the next few days we made good time as the Ucayali wound toward its reunion with the Puinahua Canal. The only events of note were a visit with the crew of a Coast Guard gunboat whose medic treated the cyst on Jeff's behind. The crew stayed with the team all the way to Iquitos, sleeping on the banks with no tents or sleeping bags and a guard on duty at all times.

Soft drinks were quite the luxury out on the river and we carefully portioned and shared the stocks, relishing the rare treat like a fine wine around campfires. One evening Ian approached John, cup in hand, to ask if he had any soda left. John said he didn't so Ian moved on, empty-handed. The next morning, he noticed John packing two large bottles of soda into his kayak hatch.

"You said you didn't have any sodas."

"These are mine." John replied.

The Marañón

"I was struck with the appearance of one, the only pretty Indian girl I have seen. She appeared to be about 13 years of age, and was the wife of one of our boatmen. It was amusing to see the slavish respect with which she waited upon the young savage (himself about 19), and the lordly indifference with which he received her attentions. She was straight as an arrow, delicately and elegantly formed, and had a free, wild, Indian look that was quite taking."
William Herndon, Marañón River, September 3, 1852

About mid-day on Oct. 20 we reached the confluence of the Ucayali and the Marañón rivers, where the waterway we had been following for nearly two months finally and formally becomes the *Río Amazonas* on Peruvian maps (the Brazilians call it the Solimões until it passes Manaus). Herndon entered the mouth of the Ucayali River at this exact spot on September 15, 1852, arriving via the Marañón to explore the lower Ucayali, of which he wrote, "*It is the longest known tributary above Brazil, and is therefore called by some the main trunk of the Amazon.*"

Smyth made the confluence on March 15, 1835 and took note of the immense breadth of the Marañón compared to the Ucayali. He camped on Omaguas Island, known now as Isla Pescadero—Fisherman's Island—across from Nauta four miles upriver of the confluence. This also was the staging area for the Great River Amazon Raft Race, where David Kelly, Carter Johnson, Mike Scales and I built our balsa log raft the day before our record-setting race to Iquitos in 2008.

Still suffering greatly under the effects of unending asthma and mosquito attacks in 1952, Che Guevara was less romantic about the confluence.

"Two more days changed nothing. The confluence of the Ucayali and the Marañón that gives birth to the earth's mightiest river has nothing transcendent about it: it is simply two masses of muddy water that unite to form one—a bit wider, maybe a bit deeper, but nothing else."

Our guardians on the Ucayali. Photo by Erich Schlegel

Jeff gets checked out by the combat medic. Photo by Erich Schlegel

The commanding officer of our guard boat, Ensign Emilio Huaco, invited us to lunch on the Navy research vessel, the *BAP Stiglich,* moored at the confluence of the Ucayali and Marañón. Knowing that this event would take longer than our usual lunch break but not wanting to decline such a gracious offer, we paddled up close to the white steel hull and asked permission to board. After a quick tour, we were ushered into the stark luxury of the air-conditioned captain's galley.

After being served cold soda, we sat in comfort, making small talk with Ensign Huaco and a few other crewmen who came and went. Then, with the subtlety of a bongo solo during a Bach concerto, Pete announced that Jason was going to take his seat in the kayak from here on out, since he had been sick for the last four days.

César, Pete and John during lunch on the Stiglich. Photo by Erich Schlegel

I suppose Pete figured he was doing everyone a favor by staying quiet about his illness over the past few days, but his announcement hit like a fart during intercourse: incredibly unwelcome but not enough to stop the progress. Pete hadn't discussed his symptoms with the team, but in the end, it didn't make much difference, since Jason had the skills and mindset to take his place without a hitch.

Privately, Jeff surmised Pete had about enough of John and was going to call it quits in Iquitos. I agreed but had a sinking feeling Pete planned on ending his expedition in Iquitos all along—which was more bothersome to me since he'd signed on for the entire Amazon after Jimmy had to leave. Oddly, after the initial shock of Pete's turd-drop, I felt pretty okay about it. The times they were a-changin'.

John switched to the stern of the tandem kayak, with Jason in the bow. Jason hadn't been privy to the efforts taken to get John in the stern and it would have been really awkward to say anything at this point. It'd only be for a day or so before we hit Iquitos so we wouldn't be slowed down all that much with John paddling out of synch.

At Porvenir the river hooked left, at the start of a 20-mile hairpin bend. The Amazon here is about a mile wide, so these curves are massive. The three kayaks held together really well at a decent pace. John synched well with Jason, as they swung double blades alongside the rest of the team. Jason added new subject matter to our discussions and it was a lot of fun to have him in the kayaks with us.

A heavy thunderstorm hit at the start of the hairpin bend. There wasn't any lightning, but the rain came in blankets with the wind and waves kicking up a bit of excitement. The water sped up with the slight narrowing of the massive river, allowing us to cruise at nine mph for several miles without much more effort. The kayaks were solid as rocks, taking side hits, climbing and diving over large waves with ease. It was a much-needed shower since I hadn't properly bathed since Contamana several hundred miles upriver. The rain, wind and waves stayed with us for about 10 miles as the Coast Guard hung close with only one poncho between six guys in their open boat.

Everyone stayed to the inside of the horseshoe bend, with the exception of John and Jason, who steered a longer course around the outside of the curve. Soon their kayak was almost a mile from the rest of us, bouncing in big waves, dodging logs and debris brought into the river by the storm and accumulating on the outside stream. I crossed the expanse of rough water and exchanged pleasantries with Jason about the storm, which was starting to subside, and then suggested that perhaps it might be somewhat shorter, safer and faster if they joined the rest of the group as we were making much easier headway on the inside of the curve. Jason, in the bow, was setting a decent pace at about 60 strokes per minute; while John, no longer nicely in synch, looked like a hummingbird on crack, doing no less than 100 strokes per minute. Jason responded that it sounded like a good idea to cut the corner and put the wind at their back, but John remained silent, staring straight ahead. After another silent 10 minutes, I said, "Well, I suppose I'm going to join the others and cut several miles off this bend."

John and Jason under watch by the Coast Guard. Photo by Erich Schlegel

At that, I turned at a 90-degree angle to their kayak and paddled straight to the pinnacle of the bend to join Jeff, Ian and the two escort boats, which were drifting without power, easily at the same speed John and Jason were working furiously to maintain. We sat and drifted along, waiting for them to join us. About 15 minutes later, they turned in our direction. Jason closed his eyes and shook his head as they approached.

It was strange and comfortable to be on a section of the Amazon I recognized from the raft race four years earlier. I knew the exact location of a cut that would save us from going around another 20-mile bend. The cut was well traveled so it wouldn't be hard to locate. Now and again we came upon well-built houses along the high banks. Many were lodges for eco-tourists, well-heeled fishermen and other folks wanting to experience the rain forest in relative comfort. The airport at Iquitos allowed easy access to what used to be one of the most isolated places on the planet.

The sun began to set well before we found the cut, making the sky a pleasant series of pastels leading into the dark horizon we faced to the east. We dug out our headlights in preparation for a night paddle to reach the lodge where we were to meet my mother, niece and our friend and fellow racer, Ginsie Stauss. César was confident he could find the way at night, since his brother managed the lodge. Pete plugged the coordinates of the lodge into his GPS before leaving the States and was also certain he could locate it, which begged the question: "Why didn't Pete bring out his GPS when John was complaining about not having one?" There was no sense in blowing on that ember at this point.

My mom, Ginsie and niece Monica spent two days flying from Texas for a morale-boosting visit. They came with much-needed supplies of food, foot powder, whiskey and snacks, which they hauled through multiple airports, layovers and hotels, and finally a four-hour pecky pecky ride from Iquitos to the remote jungle lodge. Now, all we had to do was find them ... in the jungle ... in the dark...

The entrance to the cut was precarious given the fast flow of the river compared to the slow water in the cut. Adding to the excitement was a colossal log jam right at the entrance. Water rushed through and around the jam at a tremendous pace, threatening to pin anyone who got near. All around, stumps and huge logs sprouted out like tank traps, ready to ruin a perfectly good evening. *The Best* and the Coast Guard boat overshot the cut and had to motor back up river for a safer entry.

Maw and Hinde passed through this same cut on January 25, 1828. In 1951, the young Englishman Sebastian Snow and his Peruvian guide were directed into the cut during the night, where the roar of rushing water convinced them that certain death lay ahead. Wanting to avoid such unpleasantness, they attempted to stop below a tall earthen cliff shedding clod of red clay. Fearful of the watery death they imagined just downstream, they disembarked and, grabbing handfuls of mud, worked for hours to drag their canoe back upstream against the strong current along the bank, risking burial under dirt avalanches the whole way. Before reaching the main river at the start of the cut, they were approached by two young boys in a canoe, heading into the cut, who assured the beleaguered duo that all was safe ahead and they could proceed. Laughed at by the boys and eventually laughing at themselves, they boarded their canoe and followed the kids into the void, within minutes spilling back out into the Amazon proper. Before exiting the cut, they passed the one log lodged in the middle of the river that made all the noise, and skirted it easily.

As we drifted through the cut in the dark, I set my headlamp on red and turned it backwards on my head so everyone could follow my route. While the sky was overcast, there was plenty of ambient moonlight. As a side effect of my color-blindness I had pretty good night vision, so the kayaks followed me in line and the motorboats brought up the rear. Each of us had several years' experience racing through swamps at night, so I was surprised to hear Jason describe John's anxiety level paddling in the storm and at night in otherwise perfect conditions.

Once out of the cut, César assured us the stream leading to the lodge wasn't far, so we put *The Best* out front to lead the way. Holding close to the left bank to better find the cut in the dark, we paddled under overcast starless skies. The only light was a faint glow on the northern horizon from what I figured to be the town of Tamshiyacu. The opposite shore, miles to the east, was well out of sight in the dark.

My GPS showed Tamshiyacu about 14 miles downriver. We rafted up the three kayaks and drifted as César and the Coast Guard sped back and forth, looking for the narrow side channel leading to the lodge. An hour went by and we grew concerned that we'd have to paddle back up river to the lodge, which would've been nearly impossible in the heavy current. The guys in the patrol boat did their best to find the location but had no luck. César had only been to the lodge during the day and found the thick foliage on the bank confusing. Every gap in the trees was a potential river opening.

I called Barbara on the sat phone, who checked our transponder positions and told me we'd drifted a mile or so downriver from the lodge, which she could see on Google Earth. She tried to call my mother, already at the lodge, to get someone to come out to the main river with a flashlight to guide us in but couldn't reach her.

The military guys found a few fishermen, who gave them directions to several lodges, but couldn't pinpoint any specific lodge. We devised a few alternative plans, while drifting and waiting for the support boats. There was no place to camp in the dense swampy jungle and there were no beaches. John brought up the idea of putting their tandem kayak on the patrol boat to be hauled. None of us were all that stressed, since we were perfectly safe under the eyes of *The Best* and a heavily armed military escort, so we talked and joked a bit while discussing options. Fortunately, the weather and evening were quite pleasant. I was really enjoying the rare calm moment. Jeff concluded that if the lodge couldn't be found, then we could paddle the 14 remaining miles to Tamshiyacu and sleep there. At that, John turned our inability to find the lodge into a personal affront, "Why the hell should we believe them about that distance? I've been bullshitted all night!"

Jeff snapped.

"These guys are out here at night doing their best to help us, for free! No one is bullshitting anyone! We're all doing our best to find the lodge, including the Coast Guard guys!" Just warming up, he then let John have it with both barrels, releasing a barrage of profanity and invective built up over 1,600 miles and two months of 24-hour proximity.

It felt good to have someone else deal with John, if somewhat angrily. A better leader would have stopped Jeff and pulled each aside to work out the problem, but I was out of ammo and ideas. Iquitos was the last commercial airport until Brazil and we all silently wanted John on a plane going anywhere but here. The ensuing silence seemed to reflect the team's road-weary familiarity with John. No one cared what he thought or felt anymore.

Jason did his best to put a positive spin on the moment, offering to do whatever John wanted, whether it was to paddle on to Tamshiyacu, load the kayak onto the patrol boat for a ride to town or to the lodge, or tie onto *The Best* for a tow. He said he was good for whatever John chose to do. We drifted and ate beef jerky until the two power boats joined us again to report no luck in finding the lodge.

We told them our plan to paddle down to Tamshiyacu for the night. During this brief discussion John brought his kayak alongside the patrol boat and Jason, assuming John had decided to catch a ride, climbed aboard. That gave John occasion to loudly proclaim, "Well, I can't paddle this boat by myself!"

Jason apologized and quickly turned around to get back in the kayak, but John was already scrambling onto the speedboat. With the help of the servicemen, they hoisted the heavy tandem kayak onto the launch and the two motorboats zoomed ahead to Tamshiyacu, leaving Jeff, Ian and me to a peaceful pleasant evening on the immense Amazon River. We finished with a new daily record of 87 miles, the last five or so in the company of a sweet pod of curious dolphins.

Chapter Twenty-Three
Parting Ways

Day 67, Mile 1,793

> *"Situated on a high bank on the south side of the river, distant 2,146 miles from the Atlantic; thermometer, 76 degrees. At this place the river is narrow, has only one channel, and the current is strong. It is probably the only position on the Amazon, below the mouth of the Ucayali, where vessels could be prevented from passing, up or down, by heavy guns in forts or batteries."*
>
> Captain James Henry Rochelle, 1877

We set up our mosquito-net tents on the dock of a floating gas station in Tamshiyacu, where a stray kitten fell in love with Jeff and Ian. When Ian wouldn't let it inside their tent the kitten clawed its way up the mosquito netting, curled up above their heads and went right to sawing tiny kitten logs. Jeff awoke and poked it, thinking it was another rat. The kitten took a long look at him down its short nose and curled back up until morning.

As we set up camp César finally managed to get a call through to his brother at the lodge, who reportedly would bring a launch down in the morning to visit and deliver supplies, but strongly insisted he wouldn't be bringing any of our support crew with him. A quick call to Barbara revealed that she had gotten a call through to our mother and told her of our status and whereabouts. After hanging up, I turned to the boys and assured them my mother would be here first thing in the morning. Jeff and Ian thought this was funny, since César insisted repeatedly that his brother was coming alone. The thought of my 68-year-old mom in a small wooden pecky pecky in the middle of the Amazon just wasn't in their repertoire. César assured us there was no way in hell his brother would bring anyone from our support crew, much less my mother.

A family rafting their way downriver to Iquitos. Photo by Erich Schlegel

We woke at daybreak to the sound of a barely muffled outboard pulling up alongside our floating dock. As the motor shut down, I rolled onto my elbows, cleared my eyes, yawned and croaked, "Good morning, Mom." She was standing on the dock as if she owned the joint.

Mom in Tamshiyacu. Photo by Jeff Wueste

Iquitos

Day 68, Mile 1818

"At the market I ate a piece of a grilled monkey—it looked like a naked child."

Werner Herzog

Mornings are abrupt in villages along the Amazon, and Tamshiyacu was no exception. Some Celestine prophetic connection shouts out telepathically to everyone, "Get the hell up, time's a-wastin'!" igniting a brief but intense burst of activity.

Ever the contrarian, my morning wasn't going to start until I had a couple of cups of hot joe, some oatmeal, light banter and a few goddamned minutes to figure out if I was going to have to kill someone. But nobody on that gas and beer barge cared about civil beginnings.

My mom was also an issue in need of attention. I know she flew across the equator hauling heavy resupply gear on arthritic knees (later replaced), but surely, she should appreciate my need for a moment of caffeine-fueled civility before firing up the rockets.

She appeared, as predicted, on a pecky pecky bearing the essentials of life: Twinkies, beef jerky, brownies from my sister Christine and two old friends, Jack and Daniels. Predictably, the determination of César's now somewhat sheepish brother proved lacking when pitted against my mother's resolve. I felt a little sorry for the neophyte, but he'd been warned.

We rolled out feeling unjustifiably hungover from the gasoline fumes wafting about the dock. It was really good to see my mom. After a long hug, she got right down to business. I updated her on team dynamics, and she then did her best to not take over the entire expedition.

At the upstream end of the small wooden barge were two semi-walled enclosures, each with a 10-inch diameter hole in the floor. Aside from the heads aboard the *Stiglich*, this was the most modern toilet we'd seen in a long time. After my morning constitutional, I noted a serviceman repeatedly dipping his toothbrush in the downstream side of the structure, directly in the stream of the latrine, to brush his teeth. I needed more coffee.

Before I could link together a workable number of synapses I was informed that Pete was hopping onto the patrol boat to Iquitos for some medical attention and they were firing up their engines now. I needed to get a bearing on the world and called out into the melee, "Hold on one damn minute!" I wasn't averse to any medical attention; however, there needed to be some order to this madness. I cornered Pete.

"Why do you need to go to Iquitos right this very minute, instead of waiting 20 minutes for the rest of us?"

"I don't know. They just told me they were going, so I told them I was too."

The Coast Guard guys were ready to take off. I asked them to just hang on a minute.

I drank another cup of coffee and caught up with my mom who was quickly doing her best to fulfill her natural role as the expedition leader. The guns and muscle were no longer needed, as I knew these 30 miles of river fairly well from the Great Amazon Race. We'd meet my mom and the support team at the hotel in Iquitos in two days after she and her crew completed their jungle tour.

"Okay," I told Pete. "We'll find you at the Naval hospital in Iquitos."

As Pete packed his gear he pulled Jeff aside for a quiet word of caution: "Be careful. West will get you all killed."

"I'll keep that in mind," Jeff responded diplomatically.

Pete had yet to say anything to anyone, save perhaps John, about leaving the expedition. He lumbered aboard the patrol boat with his personal gear and nary a glance back. John indicated no desire to paddle the remaining distance to Iquitos and was quickly becoming an afterthought, though his departure had yet to be verbalized. I figured there'd be time in Iquitos to lay it all out. They fired up the outboard and sped off with Pete's tandem boat loaded awkwardly on board. For a moment or two all was quiet.

We mindlessly packed gear into *The Best* and the two remaining kayaks. John loaded himself in with Erich, Jason and César and they were off.

My mom climbed back into the (her) pecky pecky and directed the driver to take her back to the lodge. In spite of my mom's lack of Spanish and his lack of English, he did as instructed. There was no point in arguing. He, too, had a mom and knew the futility of any opposition.

West sets the GPS while conferring with Jason and Pete before heading to Iquitos from Tamshiyacu. Photo by Erich Schlegel

We passed several small shipyards on the way to Iquitos, where hand-hewn wooden ship skeletons stood along the shoreline. They'd be planked and decked before the rainy season, when the river rose to float them. We just as easily could have been in Portugal or the Mediterranean 200 years ago, as skilled shipwrights turned raw timber into stout vessels without the benefit of written plans or power tools. Ships on the Amazon were built with the materials on hand, with the skills available and knowledge handed down through generations. The craftsmanship was truly impressive.

The three splits of the Amazon rejoined the main channel, which was soon augmented by the Río Nanay spilling out of the rain forest from the west. Iquitos perched at the confluence of all these waterways, which formed a huge body of water with currents pushing hither and yon against one another. We aimed for a conglomeration of moored watercraft crammed against the concrete and mud banks on the outskirts of the city. In the near distance, tallish buildings loomed and the hum of commerce and industry battled against the sounds of nature. To our right, on the opposing bank of the Nanay, military floatplanes lay in wait. In the opposite direction, farther down the Amazon, we could make out small patrol boats and river-sized warships moored to the shore.

The Best had gone ahead to arrange storage for our kayaks with the folks from AidJoy. A longtime friend of Erich's, Jonathan Shannin, ran the U.S.-based humanitarian organization. Jonathan introduced Erich to César, who guided many of AidJoy's medical aid expeditions into the jungle to help indigenous people.

Maw and Hinde stopped in Iquitos on January 26, 1828, where they obtained samples of the root from which native fishermen extracted a white solution used to paralyze fish, making them easy to scoop up. Maws' interest in the root was motivated by its potential medical uses. The two main chemicals extracted from a small group of plants are known as saponins and rotenones. Both chemicals are generally non-lethal to humans, though they can cause temporary paralysis and make a person awfully sick. With fish, the stunning effect is only temporary, so they have to be gathered quickly.

The day after Maw departed Iquitos, a drunken group of locals attacked the governor and his family, mortally wounding the politico's father and injuring his daughter. The family escaped into the jungle where they found a canoe and paddled to Pevas for help with retaliation. The drunken attackers had been provoked by changes in the tax rules requiring a portion of their produce and commodities be paid to the regional governor through the local governor of Iquitos. Having seen little, if any, return on the taxes they paid, the local citizens saw no reason why they should have to pay and, once they were well lubed, took action. Revolutions around the world typically run along similar lines.

Smyth landed in Iquitos on March 17, 1835 and recorded its population to be about 60 souls.

> *"We went first to a hut which was called the Quartel, but not having been inhabited by the human race for a long while, it was full of insects of every description, that we found it impossible to rest there till we had made a fire of rushes and leaves and burnt them out."*

Herndon stopped here on November 7, 1852, staying only one night and noting that although Iquitos was situated on a plain suitable for farming, the people preferred simply to collect whatever they needed from the surrounding forest. In his congressional report about the Amazon, Herndon related the tale of Manuel Ijurra, a member of his expedition who had spent all of 1843 overseeing a large group of men excavating gold up one of the nearby rivers. With a huge pile of gold dust loaded on a raft the size of a small village, they floated down to Pará (now Belém), Brazil. A chemist there assayed the cargo, declaring the gold pure and offering $80,000 for it on the spot. Ijurra knew he'd get much more in Europe, so he chartered a ship and took the load to Liverpool, England. There, after a year's worth of toil, expense and labor, Ijurra learned that his cargo was not gold after all. It was worthless. His story may have influenced the Jules Verne story "The Big Raft."

Men did eventually make great fortunes in Iquitos, thanks to the rubber boom. The explorer-priest Mozans (Zahm) spent time here during the onset of the boom in 1909, comparing it to Leadville, Colorado after the discovery of silver in the Rocky Mountains. *"We care nothing for politics or religion here,"* said a prominent business man to me; *"the only thing we have any interest in is the English sovereign."*

During that time, Iquitos had grown to about 15,000 residents, with many more living on rafts and ships moored at the edge of town. The harbor was packed full of boats of every make, from steamers and large sailing ships to rowboats and canoes—anything that could haul raw rubber to North America and points beyond. Farming and ranching quickly took a back seat to rubber harvesting, as a small amount of the precious sap collected deep in the jungle would sell for more than a farmer could make from a good year's crop. As goes the story of all boomtowns, raw material flowed out and money and goods flowed in. Ocean steamers unloaded luxuries and necessities formerly known to only more modern cities of the world, launching the most desolate boomtown in the world almost overnight.

Che and Alberto landed in Iquitos on June 1, 1952 and headed straight to the hospital, where they had a letter of introduction for the head doctor. Though he was absent, the staff gave the two travelers beds and meals for several days. Che's asthma was in full force, despite four injections of adrenalin per day. Perhaps if he cut down on his smoking it may have helped, though he did eliminate rice from his diet, for what it's worth. Having spent most of his time in Iquitos lying in a hospital bed Che had little to report of the town, other than a screening of the Ingrid Bergman movie *Stromboli* at the local theater.

Our lodging, at the Hotel Casona, was well apportioned, with large rooms and a pleasant courtyard complete with tiny green parakeets that immediately took to Jason, who is one of those people to whom animals seem to feel a strong kinship. Since it was early in the day, we threw the gear into our respective rooms and came together to form a plan.

I met John in the courtyard and, determined to let him take the lead on whether he would stay or go, asked whether he'd had any luck finding a partner to replace Pete for the remainder of the expedition. He had the sat phone and my mobile phone, but apparently hadn't made much effort.

"I thought Barbara was going to contact Jimmy."

"No, this is your gig," I told him.

I figured John's options were fairly non-existent, given his penchant for alienating others. For months we tried to keep John in the fold and I was still willing to have him on the team, but if John wanted to continue, it was on his shoulders.

His first call was to Jimmy where he left a detailed message. Jimmy never replied. I'm not sure John asked anyone else. Jason would have been a logical choice to take over Pete's seat, but before the subject even came up Jason made it abundantly clear that he had business back in Texas, and was flying out.

We met up an hour or two later to make plans for gathering gear and so forth. John was upset because checking on Pete's status wasn't at the top of the list. I figured Pete was being well cared for and felt pretty sure he wasn't in much worse shape than the rest of us. He blamed his downtrodden state on a local delicacy we called "little Juanies," which are similar to tamales, but made with rice and banana leaves. We all ate them and they were delicious. Pete didn't figure his daily influx of chemically-treated river water sludge into his diagnosis. Iquitos was his original cut-off point before he agreed to go the entire distance when Jimmy had to bail out. His time with John had been less than joyful and he was eager to get back to work. On top of all that, it was football season and his beloved Texas A&M Aggies were in the national championship picture. Pete was going home.

A week or two later, I learned from the cell phone and satellite phone bills that Pete and John had been calling home every single day, just to say hello and get the football scores. John's texting alone took $400 from the expedition budget. Every "LOL" and "okay" and "hi" with John's girlfriend back home cost the expedition 50 cents, and the sat phone cost $4 a minute.

We all loaded into a few mototaxis and zoomed to the Naval hospital to spring Pete. Erich, being the only truly bi-lingual team member, was allowed in while the rest of us cooled our jets outside the chain link fence, languishing in the urban atmosphere and anticipating the huge meals soon to come. Erich shortly emerged with Pete, who looked a fair bit fresher and more rested than he had when we parted ways on the fuel dock a few hours before. His newly restored look gave me a smidgen of hope that he would continue the expedition, but once back at the hotel he began making plans for the flight home.

Rest, Relaxation and Red Tape

"If you don't eat yer meat, you can't have any pudding. How can you have any pudding if you don't eat yer meat?"

Roger Waters

We were planning only one day in Iquitos, but before we could leave we had to get through customs. Barbara and my mother spent a huge amount of energy to smooth our customs experience well ahead of our arrival, to no avail. What followed was a brilliant level of nonsense that took almost four days of meetings with numerous officials, the intervention of the Peruvian military, multiple levels of bureaucracy and copious quantities of alcohol to solve. Though the Brazilian border was still some 350 miles downstream, Iquitos was our last chance to gain official permission to take our kayaks out of Peru. I found this ironic after paying thousands of dollars in official fees and opportunistic extortion to get the kayaks *into* the country—including a $5,000 deposit to guarantee that the kayaks would, in fact, leave the country. Our friends in the military insisted we complete all the paperwork to ensure the return of my deposit, which I found quaint. That money was gone.

Ensign Huaco and Erich represented the interests of the expedition in these meetings, most of which ended with me lunging across a desk to strangle some starched-shirt pencil pusher, only to be held back by my more reasonable cohorts. We were eventually granted permission to leave, but only after a particularly self-important bureaucrat inspected our kayaks and insisted on examining our paddles, which were apparently the deciding factor in his question as to whether these were, in fact, real kayaks. Admittedly, I was not the best diplomat, opting for a rather acerbic, ineffective approach more appropriate to a cage fighter. I'll forever be grateful to Erich and Ensign Huaco for their cooler heads. In the end, I had to fork over an additional $2,000 to legally paddle our kayaks out of Peru.

I can't praise the Peruvian Coast Guard officials enough. Every sailor was an utmost professional and the officers were perfect gentlemen, helping the expedition in countless ways, big and small. Ensign Huaco, Captain Bolanos and Captain Alcalá not only helped us navigate the mind-numbing customs process, they also provided detailed maps of the Amazon from Iquitos to the Brazilian border.

Remaining team members are briefed by Captain Alcala on the route to Brazil. Photo by Erich Schlegel

Pete wasn't up to joining the team and our support crew for lunch but dragged himself off his deathbed to watch the A&M game at a local bar. Having made myself familiar with downtown Iquitos during my previous visit, I was sure the Yellow Rose of Texas bar was the place to go for Texas football. Set two doors down from the famous iron Eiffel building, the bar was an embarrassment of Texas mythology, with leather saddles for bar stools, walls bedecked in garish Texas paraphernalia and waitresses spilling out of ancient skin-tight University of Texas cheerleader outfits. Pete showed up right on time for the kick off, only to find that South American *fútbol* was his only viewing option. Dejected, he shuffled back to his room to mourn. The Amazon can be especially cruel at times.

Pete, John and Jason were all departing on the same flight the next day, together with my mom, Monica and Ginsie. The rest of us took the opportunity to send six large duffel bags of unneeded gear back to Texas with them. We sorted, cussed and discussed what we needed and what we didn't. Turns out there was plenty of weight we'd been hauling for no good reason.

Jason's friends join him for morning coffee. Photo by Erich Schlegel

Several hours later the entire team and support crew, save for a heartbroken Pete, met up at the Dawn of the Amazon Cafe. The promenade, much the same as when Werner Herzog filmed *Fitzcarraldo* there 34 years earlier, is where dirtbags from

North America and Australia take great pains in affecting their world-weary, deeper-than-thou, dirty grunge looks next to locals who, in stark contrast, practice socially appropriate hygiene. Smoking their hand-rolled cigarettes, the dirtbags spread their refrigerator-door quality art on the concrete, next to the wares locals sell to earn money for food. Sitting outside on the promenade, we binged on wonderful food and drink under the paternal care of the proprietor, Bill Grimes. Bill was an expat from the American Midwest, where he'd farmed most of his life and, when time allowed, pursued the art of fishing around the world. As his trips to the Amazon became more frequent, he began to find himself more at home in Iquitos than in Iowa. After building his own two-story excursion boat—later sunk by a huge river log—he became a restaurateur. The Dawn of the Amazon Café is named in honor of his ill-fated fishing boat. He fell in love and married Marmalita and they set up shop on the Iquitos promenade, serving great food and arranging fishing excursions. Bill was a quiet, but friendly go-to man for expats in need of anything and everything on the Amazon.

West, Jeff, Bill and Ian at the Dawn of the Amazon Cafe. Photo by a kind waitress

During the light-hearted dinner, my mom asked John if he could take a couple of our massive duffel bags full of gear back with him to Texas, in order to lighten her load. She even offered to pay for the increased baggage fees, along with the airline tickets I already purchased for the two of them. Without missing a beat, John deflected the direct request for help from this great-grandmother.

"I'll have to check with Pete," he said.

Our party grew with the arrival of Ensign Huaco, César and his family, and British expat Ed Hudson. The evening was a complete joy. Ed was quite the authority on the Amazon River, and it felt a little strange keeping the secret of the new source from him. Always warm, Ensign Huaco let his crew-cut hair down a bit, though he remained the consummate gentleman.

Late next morning, as I was having coffee in my room with Ginsie and Monica, Pete stuck his head into my hotel room and said, "Okay, West, I'm out of here." John didn't say goodbye. We spent the next two hours cleaning their room, which was strewn with gear and trash. They boarded a mototaxi for the 15-minute ride to the airport, four hours before their flight and three hours before Monica, Jason, Ginsie and my mother hauled our six huge duffel bags of gear to catch the same flight.

To take on such a monumental task as paddling the Amazon River requires a great amount of confidence. Then to overcome the physical and internal conflicts inherent along the course takes a mastery of skill, endurance and grit, all of which John had in spades. I have no doubt that he could have paddled the entire route successfully; however, when working with others for a combined effort, we all must be able to work past our disappointments and constantly adjust our expectations to accommodate whatever interpersonal challenges stand in the way of our common goals. One of my greatest regrets as a leader was that I was unable to bridge the gap that lay between John, myself and the rest of the team.

After another night and day of wrangling our way through the customs quagmire, we hit the river late in the morning after a wonderful huge breakfast at Bill's café. I returned to Iquitos three years later while touring Peru with our documentary and spent some wonderful time at Bill's elbow, trading stories. He died in 2016 when his building was hit by a freak Amazon storm in the middle of the night. The outside wall collapsed, causing Bill to fall three flights to the cement below. Marmalita survived, but Bill's warm heart gave out on the way to the hospital.

Part Three
The Inland Sea

Chapter Twenty-Four
Storm Alley

Day 72, Mile 1,869

"Smooth seas do not make skillful sailors."

We moved more easily and quickly without the third boat, putting Iquitos in our wake around 9 a.m. and hitting our first white squall just after midday. It came upon us quickly, the clouds marching upstream and bursting in rapid succession, 10 rain lines in 20 minutes. The wind and waves were a welcome break from the sultry heat, and the sky stayed cloudy the remainder of the day and night, with impressive lightning bolts striking the jungle all around.

Every Amazon expedition encounters thunderstorms on this stretch from Iquitos to the Brazilian border, and ours was no exception. The heavy winds, waves of low-flying dark clouds, walls of heavy rain, incessant lightning and bone-rattling thunder come with little warning. The storms seldom last more than an hour, but out on the broad river there are precious few places to take shelter from the onslaught. Most explorers report high winds pushing them into piles of downed trees lining the banks, threatening to capsize their boats, or crush them altogether. Smooth beaches are scarce on this part of the Amazon, where the steep banks bristle with fallen tree trunks preventing safe landing.

After the storm we passed the tiny village of Pucallpa, not to be confused with the larger city up river on the Ucayali. There would be no beauty pageant here. This Pucallpa was known as New Oran when Herndon spent a night here on November 8, 1851. *"It is one of the most pleasantly situated places I have seen—on a moderate eminence with green banks shelving to the river."*

He reported that New Oran was founded when some residents of the old Oran *"found their situation uncomfortable, removed and settled here."* Herndon wasn't specific about their reasons, but it appears the original Oran continued to thrive without them. Pucallpa/New Oran looks as if it has seen better days, while Oran is a good-sized village with electricity and a water tower.

We found an entirely new world downriver of Iquitos. This entire section was dotted with villages, haciendas and houses along the left side of the river. The current was faster and the entire feeling was much more modern and less threatening.

We camped that evening on the sandy Isla Oran, about 10 miles downriver from the mouth of the Napo River, where we passed the first ocean-going vessel of the expedition, *El Manatí* (The Manatee). The massive ship seemed out of place on the Amazon, riding high on its anchor with an empty hold.

Coincidentally, in 1827, Maw met with the governor of Oran, who reported that the trade in dried manatee meat, or *vaca marina* (sea cow), was worth more than any other commodity except for metal goods. The slow-moving manatees were easy to spear as they grazed in shallow water. Their rendered fat was used for lamp oil, the demand for which all but decimated the manatee population on the Amazon River.

The next day began un-caffeinated and un-fed, as we scrambled to escape a morning storm, the wind kicking up sand, pelleting our tender skin, as we made for the water. Such an uncivilized beginning to the day left Jeff and me in foul moods until we found a scrap of muddy beach on which we could perch and brew an appropriate amount of hot coffee to properly start the day as nature intended. Ian was unruffled, the bastard.

Waves generated by the early morning storm averaged around four feet tall, straight into our faces. The storm was a lot of fun, but would have been better if I'd had my spray skirt on. In spite of the rough start, we still clocked in a healthy 77 miles for the day.

After catching up with us around mid-day the following day, Erich, César and their driver stopped in Pevas, home of the artist Francisco Grippa, an old friend of César. Grippa harvests bark from nearby trees to make his canvases, using the method local people have used for centuries to make the "tucuya" cloth. For years, Grippa split time between Pevas and Los Angeles, where he originally studied art. After art school, he spent several years in Europe studying and recreating the work of the masters. In 1972, he visited the Amazon for the first time after reading an article on the Shipibo people. He lives in a red-roofed house on the bluff overlooking the Amazon and continues to show his art around the world. Erich and César stayed there on their return trip to Iquitos, and Erich bought an original painting that might someday secure his retirement.

Pevas also claims to be the oldest European settlement on the Amazon, dating back to 1735. Isabel Godin des Odonais stopped here briefly on April 21, 1770 while traveling down the Amazon to reunite with her husband, from whom she'd been separated for more than 20 years.

202 days before, she'd departed the Spanish colonial outpost of Riobamba in present-day Ecuador with a party of 42, including her two brothers and 11-year-old nephew, four slaves, 31 indigenous people and three French hangers-on. The Indians wisely melted into the jungle when they arrived at a mission recently ravaged by smallpox. The rest muddled on for a few more days before two of the Frenchmen, including a self-styled physician named Rocha, convinced Isabel and her brothers to wait upstream while they took the party's only canoe in search of help. She agreed on the condition that her trusted slave Joachim accompany them.

Rocha had no intention of returning, and refused to help Joachim rally rescuers when they arrived at the settlement of Andoas. A month passed before the resourceful Joaquim, with no help from Rocha or the local authorities, managed to gather supplies and a handful of native people to return upriver. When Joachim arrived he found the bodies of all but Isabel, of whom he found no trace.

She was already gone, having somehow survived and wandered into a pair of locals who carried her, emaciated and half-mad, downriver to Andoas. There the authorities, who hadn't lifted a finger for Joachim, nursed her back to health and insisted she continue to Brazil in the company of Rocha, the man who had left her to die. Isabel had no words for the lowlife doctor, whom she rightfully blamed for the death of her loved ones. It was a very silent journey.

Meanwhile, Joachim paddled and trekked back up through the jungle and over the Andes to personally inform the authorities in Riobamba of the tragedy. With him he took a letter from officials in Andoas, confirming everything he was to report.

Upon reaching Riobamba, where he had lived all his life in the service of Isabel's family, he was turned away as he tried to deliver the letter to the mayor and governor of the province. After several weeks, he was arrested for being a fugitive slave. Though he presented the letter in his defense, which specified the circumstances that would grant him his freedom, the authorities were not having it. While they bemoaned the tragedy that befell Isabel and her party, they laid blame on Joaquim, failing to see any value in the fact that he risked his life and newfound freedom in the Amazon to carry the news of her presumed death 1,000 miles upstream.

Eventually, after Isabel's improbable survival and return to Andoas, word came that she was actually alive. Her first action upon regaining her strength and having the botflies removed from her skin, was to send a letter to Riobamba granting Joachim his freedom, as she had promised him in exchange for his assistance in reaching the Amazon River. Isabel continued down the Amazon and sailed to French Guyana, where she was reunited with her husband. They lived another 22 years in France and died within months of each other.

Pevas marked a significant geographic change, as the river grew in width and, after traveling generally northward for more than 1,800 miles, turned east toward the Atlantic. From here on the sun would rise in our faces and set at our backs for the remainder of the journey.

We ended the day as we started, in a huge thunderstorm. Ian spotted a beach across the river on the island of Zancudo, so we sprinted straight into wind, waves and sheets of rain to get there. We could make out the metal roof of a house on the bluff above the beach, but in this storm, we doubted anyone would come out to meet us. Erich and company landed their warm and dry speedboat in a downstream eddy on the sand beach. We pulled the boats up to high ground then waited out the storm in Erich's boat, to his chagrin. He really wanted photos of us setting up our tents in the storm, but that wasn't going to happen. Even if the tin-roofed speedboat hadn't been there we still would have sat in the rain until it stopped, to avoid getting the insides of our tents wet.

In January 1986, Joe Kane and Piotr paddled day and night through this section, racing to reach the border before Piotr's Peruvian visa expired. They became separated in a dense fog when Joe fell asleep in his cockpit. As he dozed, some men approached in a pecky pecky, took his bowline and silently began to reel him in. When Joe came to and asked what they were doing, the men didn't say a word. Finally, he swung his paddle at them and they let him go.

Lepers, Revolutionaries and Rabbis

Day 74, Mile 2,019

"Ve con dios, Peru! Te quiero!"

The next evening, we camped at San Pablo de Lorenza after paddling 73 miles; remarkably good time considering I was up most of the night with leg cramps and an overly active bladder. We were constantly searching for the right balance between hydration and electrolyte levels, both of which seemed to hover just below optimal.

I felt better after swallowing a handful of electrolyte pills and vitamin I. At lunch, Jeff mixed instant coffee with our Spiz chocolate protein powder to make a drink we called "Spaz." With the right mix of calories and caffeine we'd happened on the perfect formula for fuel. That stuff powered me all afternoon, pushing me to the 2,000-mile mark and beyond.

We passed villages settled by Jewish entrepreneurs during the rubber boom. In 1824, a group of Moroccan Jews settled in Belém, Brazil near the mouth of the Amazon, where they established a synagogue, Essel Avraham. Once the rubber boom hit, between 1880 and 1910, a growing community of Jews made their way from England, France and other European countries, traveling up the Amazon and establishing a strong presence in Iquitos. They married local women and spread out into the jungle and the eastern slopes of the Andes.

Rabino (Rabbi) Shalom Imanu El-Muyal earned a reputation as a healer among the Jews and those of other faiths; the indigenous Catholic community referred to him as "Saint Little Moses." The towns along river right in this area reflect the influence of these early Jewish settlers. Mozans, in 1912, wrote of a then-popular theory that the people of Central and South America were descendants of Middle Eastern Jews who crossed the Atlantic before Columbus. Though popular with several religious sects, this theory has been easily debunked with gene analysis.

We took the cut inside Isla Puca Playa, taking us out of the main channel and saving several miles. Toward the end of the cut, we ate lunch tethered to Erich's boat. He handed down a couple bottles of water and, in a very irritable tone, told us he was done hunting down water for us since these little towns don't sell water in bottles. That led into a long soliloquy about how this extended trip from Iquitos had been a waste of his time and money, and he would probably head back early. After he said his piece Erich didn't do more than exchange pleasantries with us, instead speaking in Spanish with César and their driver. He was tired and bored with this trip, which was certainly understandable. I wouldn't ask him to stay on; I just requested that he give us some notice so we could prepare for his departure.

When we got to San Pablo that evening we had dinner in a café then went to hunt down *helado* (ice cream) in the town square. We found some in a small *tiende,* where, to our surprise, the owner asked if we were the team of kayakers paddling the Amazon. He then told us that Rocky had stopped into his shop the previous day, asking about us. This didn't come as a complete surprise as we'd heard Rocky was on the river, heading to the Atlantic. Having accomplished his original goal of paddling the five principal Amazon headwaters, he set out to make good on his all-caps promise to National Geographic to claim the first descent of the Amazon from the new source. To do so, he passed up a $52,000 business school scholarship in order to stay in South America and press his claim, though it would come with quite an asterisk. Rather than paddling the 3,800 miles of flat water between the end of the Mantaro and the sea, he was traveling on passenger ferries. Along the way he'd been asking about us.

San Pablo had originally been settled as a leper colony, and the village clinic still treats those with Hansen's disease. Che and Alberto arrived here at 1 a.m. on June 8, 1952, aboard the riverboat *El Cisne*, on which they had traveled from Iquitos in first-class comfort for third-class fares. The travelers put their medical training to use, treating and looking after the 600 patients who lived in the colony. They stayed through Che's 24th birthday, on June 14th, which was celebrated in a gathering arranged by the director, Dr. Bresciani, and attended by Mother Margarita and Sister Alberto. Gallons of pisco were consumed and the party finally ended well past midnight.

During their stay, they played *fútbol*, had several parties and went fishing. Che even swam across the Amazon under the watchful eye of a chase boat, an adventure that took several hours and must have ended miles downriver given the swift flow. Eventually, the boys persuaded their friends to build them a raft. They christened it the *Mambo-Tango* and planned to ride it 1,000 miles downstream to Manaus, and then catch a riverboat to Venezuela. At their final party, hosted by Dr. Bresciani, a fight broke out when a drunken attendee became upset at having not been formally invited. Our travelers felt quite awkward.

> *"The episode upset us a little because the poor man, apart from being homosexual and a first-rate bore, had been very nice to us, giving us 10 soles each, bringing our total to 479 for me and 163 ½ for Alberto."*

The town of San Pablo figures in *The Green House*, the 1965 novel by Mario Vargas Llosa, who later received the Nobel Prize for literature and, incidentally, helped Piotr and his Canoandes teammates out of some political hot water in 1981. The Polish adventurers organized a street demonstration and concert in support of the upstart Solidarity movement, and the communist authorities back home were not at all amused. Friends warned them they'd be arrested if they set foot in the Polish embassy, which was a problem because their visas were about to expire. Llosa and others helped them get the documents they needed to travel overland back to the States, where they eventually gained political asylum.

As the dying sun smeared its waning colors across the heavens, we pitched our tents on the well-built floating wooden dock and did our best to fall asleep against an interminable fiesta. Lights, music and screaming kids gave me good reason to stuff earplugs deep into my ears and cover my eyes with a shirt. When we woke before dawn a boat-load of men were tied up to the dock, where they entertained themselves by staring at us as we ate breakfast and packed our gear. All at once, I was sick and tired of being the subject of someone's amusement. Perhaps it was the lack of coffee talking, but for that instant I no longer cared that we had different social mores. I couldn't give a damn that it was okay in their world to simply stare at someone different. I longed for the comfortable social norms of my homeland, where people either talk to you or leave you the hell alone. So I decided to give them a show.

With all the maturity and self-restraint of a three-year-old, I faced the crowd and let go with a long stream of morning relief into the narrow strip of water between the dock and the boat full of silent strangers. I let these guys know just what I thought of them. Served them right.

What I had not counted upon was that what I took to be a boatload of guys turned out to be a boatload of women.

Godamnit.

There wasn't much grace involved with the entrance and even less so in my exit. I did my best to nonchalantly holster-up and pretend I didn't notice the ladies, resulting in a very messy situation. Here, where Che had selflessly tended to lepers, I simply made a complete ass of myself.

Around mid-day we stopped at the village of Chimbote to check in with the navy, as directed. A sliver of Colombian territory touches the left bank of the Amazon here and follows it for some 60 miles to the Brazilian border. With three nations in such close proximity and the jungle full of smugglers, authorities in these parts are extra vigilant. The officer in charge played hard at first, demanding to see our passports and giving us the third degree. I gave him Captain Alcala's cell phone number and next thing we knew he was enthusiastically helping us along our way.

Soon after checking in at Chimbote during his 2007 Amazon descent, kayaker A.J. Rivera stopped in Caballococha, a small town on the Peruvian side. Rivera decided to paddle the Amazon when he realized he wanted more out of life than the stable career that had been the focus of his first 50 years. So he flew from Florida to Lima, hauled himself and his sea kayak over the Andes and started down the Amazon at Puerto Ocopa. To me, Rivera epitomizes the true explorer—just a regular guy who gets off his tail and decides to live.

At Caballococha he befriended some women who came to the town to teach. The previous administrator and teachers had been corrupt, and the townsfolk suspected the new ones were part of a conspiracy. During the night, as Rivera camped near his kayak, a mob assembled and attacked the women, while other townspeople gathered to defend them. One woman had her head split open, landing her in a coma. The riots were just getting started, so Rivera gave the sister of the injured woman whatever money he had and got out of town. The next few nights, as he camped on the beaches and paddled down the river, he was watched and questioned by everyone with whom he came into contact to determine whether he was part of the conspiracy.

We camped on Isla Loreto, on the Colombian side of the river about 35 miles upstream of the Brazilian border. César got a huge fire going, but the mosquitoes remained, unabated. That evening, our last in Peru, Erich announced he would turn around before the border. He was ready to go and we didn't blame him. He'd been a tremendous help to the expedition and we would all miss him. At dawn we said our goodbyes to Erich and César, and they motored upriver as we paddled into the rising sun.

Chapter Twenty-Five
You, Again?

Day 76, Mile 2,121

> *"Hello… Newman."*
> Jerry Seinfeld

We reached the border around mid-day, following the contour of a low sandy island forming the easternmost scrap of Peruvian territory. The sleepy outpost of Santa Rosa occupied the swampy tip of the island, across the strait from the more imposing twin cities of Leticia, Colombia and Tabatinga, Brazil.

We paddled straight to the Peruvian Coast Guard barge moored along the modest waterfront, where Lt. Carlos Avila greeted us warmly. We quickly tied the two kayaks to the barge and began exchanging pleasantries. As we chatted my solo boat came loose and started swiftly downriver without me. I dove in and recovered the floating bow line before it got too far. A gaggle of helpful crewmen pulled me back aboard and we repositioned the kayaks between the barge and the shoreline, out of the current.

Lt. Avila spoke perfect English and wanted to hear of our journey. Lunch was laid out within minutes on a table under the shade on deck and we enjoyed soft drinks and spaghetti in the light breeze while recounting the highlights of the last 2,000 miles. The lieutenant was a close friend and colleague of the officers we met in Iquitos, who briefed him about our expedition.

It was a welcome jolt from the endless slog through the wilderness to dining with real plates and ice-cold sodas. The polite reception and gracious company served as an unambiguous contrast to the storms, gunmen, poor nutrition and generally squalid conditions along the river, even though such experiences held their own prehistoric attraction. With unusual restraint, I stuck to a single helping, though I could've easily put away an entire pot of noodles on my own, with room to spare.

Lt. Avila laid out the day's plan: get our paperwork processed through Peruvian Customs, then take the patrol boat across the river to Tabatinga, to meet with Brazilian Customs. After we cleared customs, Lt. Avila would personally introduce us to the regional commander of the Brazilian Coast Guard. With plates cleared and orders issued all around, he walked us down the gangplank to the dirt path that led a half-mile into the town of Santa Rosa. We met the customs guy at an outdoor café, with Lt. Avila explaining everything as we presented our paperwork. The customs guy disappeared into one of the buildings, and Lt. Avila led us to the local head of the National Police, who ordered one of his men to fetch us some cold Inca Cola, then stamped our passports with a flourish. Lt. Avila then brought us to the immigration guy, who stamped our passports again. Then we doubled back to the customs guy, who had finished our boat paperwork. We were amazed at the gracious help provided when we had nothing to give in return but our earnest appreciation. The efficiency with which the officials worked in this sparsely populated outpost made a sharp contrast to the hellish bureaucracy we endured in Iquitos.

We did our best to match the lieutenant's purposeful pace over the dusty pavement glaring in the equatorial sun. The air always seemed a little hotter and more stagnant when we left the river, and though the cold drinks had boosted our energy level we felt a strong urge to hunker in one of the many grass-roofed pavilions, under which savvy locals sought refuge from the mid-day heat with the aid of adult beverages and tropical music. We had just one more door to darken before heading back to the Coast Guard barge, but it wasn't going to be that easy. I spotted the familiar pale figure from 50 yards away. As he walked quickly up the empty street, squinting in the harsh sun, I thought of many different ways to address him, but as we approached one another I opted for my tried and true defense against my overtly aggressive first impulses: overwhelming passivity. A passivity that made Gandhi look like a bully.

"Hi, Rocky." We shook hands.

I wanted to punch him in the face.

He filled my self-imposed verbal voids with non-stop chatter about how he'd been looking for us. He asked all kinds of questions, and I kept my replies as short and vague as possible. I introduced him all around, but Lt. Avila was in no mood for socializing and barely slowed his pace. Jeff, who can normally be relied upon to complete whatever social colloquy is required, remained fairly mute, as did Ian. Awkwardness added to the humidity, making the air even thicker. The three of us ran behind Lt. Avila and Rocky kept pace. At Rocky's request, we stopped just long enough for Lt. Avila to take a picture of the four of us. My tiny headband camera was rolling. I figured the video record was incentive for me to behave.

Unease hung thick, as Rocky stumbled through his attempts to engage us, while we received the last of the paperwork required to transport the kayaks out of the country. Ian and Jeff responded to Rocky's attempts at conversation with the enthusiasm of a blind date staring at her watch. Even after bad-mouthing me to National Geographic, to my teammates and sponsors, in *Outside* magazine and on various websites, then declaring that he was going to beat me down the Amazon and despite having been politely but unequivocally booted off our team back in the Andes, he still believed we wanted anything to do with him.

He followed quick on our heels, shifting focus to each of us in turn when the others answered in monosyllables or not at all. At one point, he asked for a ride back across the river to Leticia, in the patrol boat. At first, I ignored the request. When he repeated it, I told him it wasn't up to me. Lt. Avila was visibly annoyed, but professional. His time was valuable and did not include socializing. Without delay, once all the paperwork was completed, we followed his quick pace down the single-track dirt path through the reed fields back to the barge. All the while Rocky droned on in his signature high-pitched monotone, telling about his ordeals along the river and peppering us with questions. He fell in line with us behind Lt. Avila and asked where we were staying that night. Our non-committal answer didn't dissuade him in the least, as he offered suggestions and plans to meet up before continuing down river.

"We're meeting with the military and aren't sure what they have planned for us," I responded, an answer I repeated many times in various forms. It had the benefit of being factual, though incomplete.

At the barge, we marched across the bouncy wooden plank onto the deck, as servicemen leapt to attention, saluting their commanding officer. Lt. Avila calmly issued orders on various subjects, and his men jumped to comply. As we were herded aboard the large heavily horse powered, semi-inflatable military patrol boat, I whispered to Lt. Avila that Rocky wasn't with us and wasn't our friend. Apparently, he picked up on this fact pretty early on. Lt. Avila quickly told Rocky, in English, that there was only room for three on the boat, though the boat could easily accommodate 20.

As his men cast off the mooring lines, Lt. Avila called out in Spanish something to the effect of, "Get that son of a bitch off my boat!" The good officer had no idea Rocky was fluent in Spanish. Without hesitation, a handful of servicemen hustled Rocky up the gangplank. Perhaps we got a bit more enjoyment out of this spectacle than we should have, joking amongst ourselves whether we should have Rocky cavity-searched and thrown in the brig for a few days. We just wanted him out of our hair.

When we got to the Brazilian side, Lt. Avila once again ushered us through a maze of red tape, this time with the formidable Brazilian bureaucracies. The immigration police asked a couple of questions about our expedition, then stamped our passports with visas allowing us three months in Brazil. We then went to the Brazilian naval base to meet with Cmdr. Robson Neves Fernandes, who controlled every inch of the Amazon down to Manaus. His special assistant looked exactly like President Barack Obama and we told him so, which drew a big laugh.

"Everyone says that," he reported with that classic smile.

Cmdr. Fernandes was very interested in our expedition and, in perfect English, asked us to email him our coordinates daily, as Barbara had done with the Peruvian Coast Guard. Then, accompanied by a full contingent of military personnel in spotless well-pressed uniforms that made us look even worse than the scruffy dogs we were, we piled into a military van and drove to the dock, where we all boarded a large patrol boat and motored across to the Peruvian barge. Lt. Avila drove his own patrol boat back across.

Salutes all around as officers and men from both countries addressed each other formally, and the three of us bedraggled slugs stood around looking like bedraggled slugs… bedraggling. As he inspected our kayaks, we answered Cmdr. Fernandes' questions and described the communication and navigation gear we carried.

The commander then surprised everyone by attempting to climb into my kayak in his jumpsuit and military boots, nearly tumping over in the process. Using the size of his boots as a damn fine reason to abort the attempt, he graciously reclaimed the dock as his entourage traded anxious glances. He was concerned about the turbulent waters we would encounter, and we assured him we could handle any conditions in our kayaks. The commander then warned us about pirates and told us not to paddle at night. We did our best to respectfully explain we'd already come through almost 2,000 miles of rough water, pirates, drug dealers and all kinds of other hell, so Brazil wasn't exactly our first rodeo. Cmdr. Fernandes nodded politely and said he would ask his superiors for permission to allow us to continue into Brazil. We were to meet him the next morning to hear the verdict. Instantly Ian, Jeff and I were figuring out contingency plans in case they denied our entrance. The permission would be greatly appreciated, but merely a formality in our books. We were going to the Atlantic.

Mickey, Ian, Lucy, Jeff and West in Leticia. Photo by a kind bystander

Lt. Avila took us back across to Leticia, showed us to a nice hotel and graciously accepted our invitation to supper. After bidding Lt. Avila goodnight, we ran into Mickey Grossman, who was sitting with a really pretty gal in a corner café. She wore the local uniform: mini skirt and skin-tight tank top, which garnered no complaints. Lucy, whose smile launched a thousand kayaks, joined Mickey's trans-continental hiking team after meeting him in Quito while on vacation from her home in Argentina. Her English was perfect and she calmly tolerated Mickey's excessive energy level, in the quiet way many intelligent women sometimes indulge the fruitless excesses of their male counterparts. Mickey was an Israeli-American and a former special forces operative for the Israeli army. After surviving cancer, he determined to cross South America at its widest point, from Ecuador to Brazil, borders and visas be damned. He'd been jailed several times and held hostage by terrorists, escaping each time with determination, fortitude and a singular ability to irritate authorities, which I found particularly charming. Mickey and I became acquainted while planning our respective expeditions, occasionally kidding about a chance encounter, which is exactly what happened.

Leticia was clean with pretty streets and quaint shops. It's a duty-free town with plenty of high-dollar stores. The traffic is orderly, restaurants clean and garbage service efficient. Trees line the boulevards and the people were really pleasant. It was easy to feel like a high-roller, especially when dealing out Colombian pesos. Our hotel room was a cool 100,000 *pesos*, about $33.

The Land of Burning Embers

"BRAZA" (burning embers) is a word found in the Spanish language as far back as the 12th century. It has been used to make the word "brazil," as descriptive of certain woods that yield a reddish dye. From this has come the name "Brazil," given to that vast district of South America which is crossed by the equator, and in which these products are so frequently met with. In very early days, the woods were the object of considerable trade. Although correctly called "ibirapitunga," from the place of production, the name of "brazil" stuck to them, and it has become that of the country, which seems like an immense heap of embers lighted by the rays of the tropical sun.

Jules Verne, 1881, *Eight Hundred Leagues on the Amazon*

In Leticia, motorcycle helmets are mandatory, except for infants and children, who can be slung over an arm or perched on the gas tank. We enjoyed a pleasant stroll over to Tabatinga, the border between Brazil and Colombia being almost non-existent. Tabatinga resembles a comfortable south Texas town. It was somewhat surreal to see the Halloween ornaments in store windows and decorations around town, along with kids wandering the streets in costumes. Prior to this, I thought such trappings were purely North American.

We arrived early for the meeting with Cmdr. Fernandes, who immediately eased our worried minds and granted us permission to continue into Brazil, then presented us with detailed maps of the river all the way to Manaus. Rolling the charts out on the table, he recommended routes to take around various islands and strongly suggested we visit the beautiful city of Tefé. He would call ahead to naval stations along the river with orders to accommodate us and help in any way possible. The red carpet was laid out and we were thankful to have it. If Cmdr. Fernandes were a hugging man, which he emphatically was not, I would've hugged him.

Getting briefed on the route by Commander Fernandes. Photo by West Hansen

Boats of any size are rarely allowed to pass from Peru into Brazil. This state of affairs reaches all the way back to 1494, when Pope Alexander VI coerced Spain and Portugal into dividing the New World in half along a line that very roughly corresponds to the modern border of Brazil and Peru.

Control of trade is all about pride and national sovereignty, which plays out in very pragmatic ways between the towns of Santa Rosa, Leticia and Tabatinga. Locally, no one pays the three bordering countries much mind, but when something official comes up, suddenly everyone is by the book.

In 1850, a circus troop made its way down the Amazon on a huge raft built to carry all the animals and performers. At the border, the Brazilian commandant famously required the circus to abandon the sturdy raft and build an entirely new one using timber cut from the Brazilian side of the border.

In 2010, British adventurers Patrick Hutton and Jason Varndell built a raft over two large wooden canoes and steered it from the Urubamba down into the Peruvian Amazon, intending to continue all the way to the Atlantic. Though the boys had successfully navigated almost 1,500 miles, Brazilian authorities deemed their raft unsafe and refused to allow it into Brazil. Hutton and Varndell considered sneaking across the border under the cloak of darkness but thought better of it after learning of two police checkpoints 20 miles downstream. The penalty, if they were caught, was $20,000, possible jail time, deportation and a ban from entering Brazil for 20 years. Heartbroken, Hutton and Varndell abandoned their dream of Huck-Finning the Amazon and continued to the Atlantic on public riverboats.

William Smyth stopped in Tabatinga on March 24, 1835, noting that the fort was entirely in ruins, with just four brass guns missing their carriages. The garrison of Ticuna Indians numbered between 400 to 600, the count being imprecise because, in the estimation of the English lieutenant, "they were a restless and wandering race." The Brazilian commandant had strict orders to allow no boat to pass without a thorough inspection. Smyth came ashore and as the commandant's men searched his cargo, the padre, whom Smyth had met upriver, floated by on a heavily laden raft. Shots were fired at and around the raft as it drifted, but the ungainly craft stood no chance of returning upriver against the current. Shrugs were exchanged all around.

In all likelihood many hundreds of rafts have simply drifted by the command posts at the border, as the glorified piles of logs tend to be good at floating, but propulsion and steerage leave a lot to be desired. In a letter to his mother, Che recounts that both he and Alberto slept as they drifted past the checkpoint. Once they noticed the lights of the city, it was too late to effect any change in the direction of their raft. The next day they caught a ride in a canoe back to the border. In Leticia they took jobs coaching a local soccer team to earn money for airfare to Bogotá. The squad was so bad that Che and Alberto ended up playing, and led the team to the finals for the first time ever. During the flight to Bogotá, Alberto regaled their fellow passengers with tales of an earlier flight over the Atlantic Ocean during which three of the four engines gave out. The detail of this soliloquy was so convincing even Che got goosebumps, though he knew of course the flight took place only in Alberto's imagination.

Horizon Lines on the Solimões

Feijoal: Day 77, Mile 2,156

> *"The padre had some large land tortoises, which were considered good to eat, and even preferred by some people to tartaruga, or the water tortoises. I cannot say I admired such diet, but in a country where monkeys and vaca marina are considered delicacies, and snakes and alligators have been eaten, not to mention human flesh, anything will go down."*
>
> Henry Lister Maw, 1828

After our meeting with Commander Fernandes and brunch with Mickey Grossman and his impeccably uniformed team of hikers, we were extremely happy to be on our way without any further delays. The tender trappings of city life, though pleasant, no longer felt as homey as the river and a nice beach camp. The landscape transitioned abruptly after we cleared the large islands at the border. What the day before had seemed a really massive river was now a quaint memory, as the Amazon, now called the Rio Solimões down to Manaus, grew even more immense.

The houses changed to bright, whitewashed New England clones, high on the dirt and limestone cliffs of the northern bank, blanketed in the omnipresent green. From a distance, we could have been paddling along the coast of Maine. Under a cloudless blue sky, we discussed and adjusted our course throughout the afternoon while learning where the mass of water preferred its strongest flow.

Despite the new political jurisdiction and name of our liquid highway, it remained the most voluminous river in the world, and we continued on its broad shoulders toward the sea. Despite leaving at 1 p.m., we covered 35 miles and made our first Brazilian camp on a mud beach at Feijoal, where we were immediately swarmed by kids and teens.

We made a habit of befriending kids when arriving in towns. They were almost always the first to greet us, and showing a little friendly interaction with children naturally caused the adults to relax. When my own daughter was young, I always felt affection toward people who connected with her in that universal language of childhood. Mozans noted the same when visiting villages along the Paranapura River in 1908: *"You like my child, I like you."* He found that after a child accepted his attention and trust, the family was soon to follow.

Ian and the Feijoal welcome committee. Photo by Jeff Wueste

Smiles and friendly gestures were about all we had to go on. Prior to the expedition I studied some Portuguese phrases, but once in Brazil I realized that though written Portuguese is very similar to Spanish, the pronunciations are completely different.

Late into the night, under the almost full moon, three teenage girls pressed their noses against my tent screen, whispering and giggling while talking to me in Portuguese. I got out my translation dictionary and told them to go away several times but was met with indifference. Finally, they ran off when I brought out the video camera. I ignored people who came down to inspect our boats in the dark until they sat on the fragile decks of the kayaks, whereupon I leafed through the phrasebook and told them to get lost. It was a hell of a way to learn Portuguese.

Ian's watch alarm went off at 4 a.m., just in time for all the folks to come down to the river en masse. I'm not sure if they were going fishing or just had a calling, but it was time for us to get going. For the remainder of the journey, in order to get a decent night's sleep and to reduce the problems inherent with polite travel habits, we avoided towns and villages.

West being swarmed. Photo by Jeff Wueste

Around midway through our 78-mile day, Cmdr. Fernandes passed us in his Godzilla-sized battle cruiser, its engine rumbling like thunder over a volcano. I swear, the water shook for miles around that marine gray goliath. The commander was on the upper deck with binoculars and camera, waving and giving us an enthusiastic thumbs-up. Painted on the vessel's side in huge block letters were the words *Comandante do Porto*—"Commander of the Port," which spoke volumes. The enormous block of floating steel didn't move fast, rather plowing through the water at its own lumbering pace. Shortly after our break the following morning Cmdr. Fernandes rumbled by again, shouting encouragement and giving us another thumbs up. His presence, though intermittent, gave us confidence that we weren't so alone. Pirates in these parts are noted for their opportunism more than their prowess. With such firepower nearby, perhaps the black hats wouldn't find the opportunity to approach us.

The straight sections of river were now so long we couldn't see past the curvature of the earth. Despite its oceanic dimensions, the river was hospitable. We experienced no real cross currents, wind or waves, just good paddling conditions in huge water. We read that the Amazon was supposed to get pretty rough below Manaus, but for the time being we enjoyed really pleasant conditions.

Commander Fernandes' battle cruiser. Photo by Jeff Wueste

On this section of river in 2007, a pecky pecky full of petty pirates rammed A.J. Rivera's plastic kayak. One of them grabbed his bow line and started to reel him in; Rivera cut the rope with his machete and started to paddle away. Brandishing their own machetes, the pirates grabbed his stern rope and dragged him to shore. The pirates identified themselves as members of the Tikuna tribe and spoke Spanish, which meant that they were certainly not members of the Tikuna tribe and didn't normally speak Spanish.

After rifling through his gear, looking for drugs and tasting the various drink powders for evidence, they stole Rivera's decoy money in his decoy wallet. They didn't find his stash of U.S. dollars and local cash, but they did get his only credit card. Temporarily despondent (all successful Amazon kayakers become temporarily despondent; the permanently despondent rarely get far), Rivera spent the night in a cheap hotel in town, then caught a ride back to Tabatinga for a replacement credit card.

Few expeditions have passed through this section of the Amazon without similar tales to tell. The region is rife with bandits and stories of weapons, hold-ups, thievery and attacks. Joe and Piotr carried machetes and stared down a few barrels in these parts. Such was their apprehension that one dark night Piotr lunged at a couple of would-be attackers with his machete, only to find the stalkers to be their lifejackets drying in the breeze. We were the exception in this stretch, blissfully left alone to enjoy the ever-widening river.

Ian and Jeff did their best to throttle down to my slower pace, but every so often would find themselves a few hundred yards ahead. My shoulder was weak, with the pain increasing throughout each day, and my energy level was waning after almost 2,500 miles of paddling. A two-man boat is simply faster than a single, and the fatigue of keeping up was wearing me thin. I could tell that Jeff and Ian were frustrated with my slow pace but they were too good to complain. Instead they offered to take more of my gear, though there was really nothing left that weighed all that much.

Jeff and I have a long history together, but Ian was still a bit of an unknown. Until Iquitos I'd spent most of my attention oiling the squeaky wheels of the expedition and paid little attention to the steady Ian. Now that it was just the three of us I had more time to talk with him and he turned out to be a pretty good egg with a levelheaded temper and dry sense of humor that I greatly admire.

We spent our time on the water with music trivia and conversations that had nothing to do with paddling the Amazon. Gilligan's Island was a recurring theme, with Jeff and I consumed with romantic thoughts of Mary Ann. We agreed the professor was an idiot for not putting all his brainpower into pitching woo with her. Due to his age difference, Ian had little to contribute to this academic subject, for though he'd heard of the show he was wholly ignorant of its nuances, or even the eternal schoolboy debate of Ginger vs. Mary Ann.

Navigation and pace were constant matters of discussion. The river was miles wide and the islands were huge, so it was no small task to plan well ahead for the shortest route. Our two daily breaks became more efficient as we worked to keep them under 30 minutes apiece. Even when we were moving slowly, we were still moving. Fortunately, all were on the same page with this philosophy. Jeff was slow as a turtle in the morning, but we adapted.

We were a bit over halfway through the total distance of the Amazon River now, though we didn't take the time to let that sink in. With plenty of water between us and the Atlantic, we took each mile and each minute as it came and hoped to finish by November 25, which would be 100 days since the start of the expedition.

A typical sand island in the middle of the mighty Amazon. Photo by Jeff Wueste

Chapter Twenty-Six
The Routine

Day 79, Mile 2,314

"A routine negates the act of having to will or motivate yourself to do something. Willpower is finite and motivation is not constant."

Tri

On October 31 we covered 78 miles, which landed us about 273 miles inside the Brazilian border since Tabatinga the Amazon had been on a west-northwesterly tack toward the equator. River traffic was common, as were good-sized towns on both sides, but the ever-widening expanse of water left plenty of room for us to disappear in the scenery. About 20 miles downriver of Tonantins, we found our usual pristine beach campsite, far from any community. Aside from a bag of Halloween candy Lizet sent down with the support team and the apropos but commonplace appearance of a swarm of bats, this Halloween unfolded like most other days on the river. It was the 79th day of the expedition, and we were comfortable creatures of habit.

Ian's wristwatch alarm woke us at 4:30 each morning, within a few seconds of my own. That was my signal to sit up and stuff Debbie Richardson's sleeping bag into the yellow drybag, then put the clothes I use as a pillow into the blue drybag. Our racing friend, Debbie, had been kind enough to lend her expensive down feather bag to my daughter to use in the Andes, though she hadn't counted on me absconding with it all the way down the Amazon. I then packed all the electronics into the small yellow drybag. Each time I prepped a bag for loading into the kayak, I stacked it next to the door of the tent. This assured that I didn't forget anything in the darkness before sunrise and coffee. I immediately loaded the sleeping bag, clothes, electronics and toilet kit into the back hatch of the boat.

The computer, in its protective case, went into the tandem's middle hatch. I put my loose things—toothbrush, malaria pills, hunting knife and other sundries—into the tiny hatch behind my cockpit. Delaying my morning pee until all was packed helped me hurry through the routine. Then, after most everything was packed, I donned the damp blue shorts and green shirt I wore every day for more than two months.

Our standard campsite. Photo by Jeff Wueste

While I did all of this packing, Ian heated water for our instant oatmeal and coffee. One pot of water rendered each of us a cup of oatmeal (two packs each) and two cups o' joe for Jeff and me. None of us cared for the blueberry oatmeal. We were okay with the apple cinnamon, but really coveted the brown sugar cinnamon flavor. Such minutiae gained an enormous amount of importance with every mile.

I made my coffee in the same cup in which I mixed and ate the oatmeal. The last oatmeal flakes went down the hatch with the coffee, so I only had to wipe the cup clean and store it away for the next meal. Blessed with youth and big biceps, Ian pumped river water through the purifying filter, and I took over collecting and storing our trash, of which there was a surprisingly large amount. Jeff covered all other areas. Not being a morning person, we gave him a bit more time in the sack and handed him his first cup of joe as he sat in his tent to jump start whatever neurons were functioning in the pre-dawn darkness. There was a certain harmony in these chores, as no one sees others slacking and we all did something for the group. There really wasn't much to say, so we normally kept a comfortable silence.

The last thing we did was pack up the tents, which we had down to an art form. Before taking down the tents, we picked each of them up and gave it a good shake to rid the innards of sand, bugs and debris. We helped one another, so it all took but a minute.

I'll take my drink to go, thank you. Photo by Jeff Wueste

The evolution of a routine and simple chores came naturally, out of necessity, personal preferences, individual and shared values. I credit the success of our expedition in great part to these routines, which became a source of comfort in an otherwise unpredictable endeavor.

We packed everything, secured the hatches, then dragged the kayaks down to the water to shove off. All of this, from Ian's alarm to the first paddle stroke, took about an hour and 45 minutes. Forcing the routine only caused problems and irritability, so we all agreed to get up earlier each day until we arrived at the magic hour of 4:30 a.m., which put us on the water just before sunrise.

The next day we woke up and did it all again, pushing on until after dark to notch an astounding 85.6 miles. The lights of Foz do Jutaí twinkled two miles away on the opposite side of the river, as we pressed on with an escort of blue and pink dolphins dancing in our wakes. The pinks were uncharacteristically active, following us for several miles. The evening was calm, though finding a camp spot was nerve-wracking. In the darkness, it was near impossible to see where the river stopped and the shore began. Our reference points were out of whack and in the darkness the current seemed to speed up alarmingly. I checked the GPS frequently and though it assured me we were traveling no faster than usual, I couldn't escape the out-of-control feeling of being swept downstream.

Eventually, we spotted a small spit of sand on the upstream side of an island just off the left bank. Landing was dicey, as the area near the bank was crowded with the branches of downed trees threatening to tump us over or grab our rudders. Finally, under the reflector gaze of caiman eyes, we landed and quickly dragged ashore.

Normally beaching about an hour before dark gave us ample time to perform camp duties, eat and hit the sack before the bell sounded for the mosquitoes to start their shift. Smelling fresh blood after dark, they immediately lit upon us in desperate swarms. We hustled to light a huge fire to ward them away, to little avail. Our only recourse was to eat quickly and get into our tents. Jeff and Ian's routine included a few minutes of mosquito killing as soon as they crawled into their tent, while I opted for a more clandestine entry into my tent, to keep the pests from entering from the get-go. This hustle at the day's end wore on us much more than we anticipated. We really needed that time at the end of each day to decompress.

The next day we paddled 74.1 miles and, having learned our lesson, finished well before darkness fell. The current was slow and the sun beat us down all day with no breeze to speak of. We were tempted to find some shade and sleep the afternoon away, but that's not how you complete an expedition, so we slogged on.

I downed a double-dose of Spaz and fueled by the extra protein, carbs and caffeine I felt strong all afternoon. I even slowed down for Jeff and Ian now and then, which was quite a role reversal. Steady calorie intake was critical to our success and took on a standard regimen: two packs of oatmeal with a cup of joe in the morning, Spaz around 8:30, nibble on snacks while drinking a quart of flavored water through mid-morning, lunch around noon, another Spaz with some more substantial food at mid-afternoon, and a freeze-dried dinner in camp.

During our afternoon break I enjoyed a can of tuna con Cholula sauce with Ian, who had developed a taste for the same delicacy. Protein was in short supply, making canned tuna a necessary staple. Topics for the day included independent artsy films, music trivia and the elusive mystique of the feminine mind, which exists beyond the comprehension of our testosterone-infused brains.

Ian took responsibility for navigation, studying the maps and estimating our daily time and distance. By his reckoning we would finish the expedition by the last day of November, six days after my contrived 100-day goal, which was still possible if we pushed hard. So when Ian suggested a stopover in the town of Tefé, I voted no, in order to keep us moving. When Ian said he wanted to stop so he could video chat with his three-year-old daughter Addie, we all agreed to stop in Tefé. I mean, a daughter aces everything, so there's simply no countersuit.

The next evening, after dark in camp, we saw a motorboat coming our way, moving fast with two large spotlights scanning us and the river. I stabbed my hunting knife into the sand next to my tent door and did the same with the machete next to my boat, then the three of us stood and waited to see what was going to happen. We were totally exposed on a sandy beach, with a half-mile or more of sand between us and the jungle. The small motorboat landed about 20 yards downstream, and six guys got out and approached. I didn't see any weapons, and I started with what little Portuguese I knew, poorly pronouncing, *"Boa noite!"* which simultaneously conveyed our peaceful intentions and the fact we don't speak Portuguese.

Communication was tough, with the three of us lapsing into Spanish way too easily, so I finally brought out the translation dictionary. The men worked at a wildlife refuge and recently had some problems with poachers in the area, so their diligence was understandable.

We didn't know until a few days later that we had inadvertently camped in a protected wildlife area, the Estação Ecológica Mamirauá. Established in 1996, it was the first such reserve in the Brazilian state of Amazonas. It was created to protect a vulnerable subspecies of monkey, the white bald-headed uakari, but also had the effect of nurturing a complex ecosystem with hundreds of species of plants and wildlife. The word *mamirauá* is best translated as the young calf of a river manatee. It is also the name of the large lake located in the center of the reserve.

I explained our expedition to them and promised we'd be gone at first light. They gave us permission to camp, and the mood eased up pretty quickly. A few of the men wore T-shirts from conservation organizations and recognized the "Project Amazonas" decals on our kayaks. Jeff took pictures with all of them, which seemed to be a real icebreaker. He then gave a tour of his boat and the workings and they were particularly impressed with our carbon fiber paddles. Pleasant as they were, they stayed around well past our bedtime, in spite of our hints, like donning our footy-jammies, putting out the cat, smearing on face cream, starting the dishwasher and turning out the porch light. Finally, I thanked them and told them in Portuguese that we were going to sleep, so we all shook hands and they shoved off.

Hovering unnoticed just outside the perimeter of our gathering were the permanent residents of the beach that we had inadvertently displaced. Just as soon as we zipped into our nylon cocoons, hundreds of sea gulls descended upon us out of the darkness. I don't believe they are normally nocturnal, but for us they made an exception. Circling our tents on wing and claw, they commenced to squawking at the top of their lungs and did not relent until we crawled back out, quite bleary eyed, for breakfast. A mere inch away from my scalp, they crowded against the screen walls bellowing their disregard for our presence. The foam earplugs did their best but weren't built for this type of decibel level. More than once, we leapt up in frustration to rush out and scatter the squadron of protesters. At one point, I ran naked up and down the beach, wildly swinging my machete in hopes of nailing one of the bastards, if only to set an example for the rest. Jesus, I was tired. When the alarm finally sounded I skulked out of my tent and absently began the packing ritual. The three of us looked like the morning after remnants of a frat party, offering single syllable expletives instead of the usual salutations.

Tefe

Day 84, Mile 2,690

> *"But if you're gonna dine with them cannibals Sooner or later, darling, you're gonna get eaten..."*
>
> Nick Cave

Nearly 70 miles of paddling brought us to Tefé, which was a far cry from the quaint little burg Cmdr. Fernandes described. It's a sprawling jungle city with little regard for traffic laws, lots of mangled construction and a harbor crowded with floating houses and shacks. The din of the bustling waterfront was a stark contrast to the subdued sounds of the river and jungle.

West introducing the team to the Coast Guard in Tefe. Photo by Jeff Wueste

Upon entering the large harbor, it became clear that we would have trouble locating the small hotel, Pousada Multicultura, amidst the crowded cityscape, so we flagged down a passing Navy patrol boat for directions and followed them to their official barge, where they rustled up an English-speaking sailor. Barbara, ever diligent and reliable, contacted the Navy before our arrival, so they were expecting us. A beautiful officer named Ivana escorted us into town to meet the Captain, in spite of the fact that we were dog-tired and stunk to high heaven. The young Captain was incredibly gracious and sent a driver to take us to supper, after stopping at the hotel to shower.

Prior to dinner, we visited with the hotel owner, Betina, and her assistant Noella, both from Denmark. Noella, the consummate young world traveler, joined us for dinner to help translate. That evening Betina wrote out some Portuguese phrases we routinely needed out on the river: who we were and what we were doing; requesting permission to camp; where we could find food. She also bought two dozen eggs for our morning meal, which isn't part of the typical Brazilian breakfast.

During his stay here in 1827, the English lieutenant Henry Lister Maw wrote extensively about slavery and cannibalism, both of which were practiced quite openly in the region. The Brazilian government at the time allowed white settlers to capture indigenous people and enslave them for 10 years. After that time, it was assumed the slaves would have been "civilized" and therefore were required to be set free. Of course, most slaveholders ignored that detail and the captive people, having no legal recourse, remained enslaved.

Reports of ritualistic cannibalism existed for centuries in the jungles of present-day Peru, Venezuela and Ecuador; however, what Maw reported of the practice in the jungles around Tefé was altogether different. To escape the slavers, many indigenous people fled their ancestral lands along the Amazon River, where fish, manatee and turtles were plentiful. Deeper into the jungle they found little nourishment and cannibalism evolved for practical rather than ritualistic reasons. Tribes resorted to capturing people from other groups, then either sold them off to slave hunters or ate them.

According to Maw's account, these captives were not mistreated. When it came time to kill for food, a poison dart was shot from a distance and the victim was dragged away for butchering. In a state of learned helplessness, the remaining captives did not fight and allowed the fallen to be taken without a struggle. Maw repeated a story told to him by a man whose father was in the jungle on a trade mission. He landed in an encampment where he traded his goods, and as is custom was offered a bowl of stew. At the bottom of his dish, he found a thumb.

Maw related another story about a young woman in one of these enclosures. When a slave trader tried to purchase her she refused, preferring to stay with her family and be eaten rather than be enslaved. Stories like these were met with revulsion and fascination in Europe and North America, though the invading cultures took no responsibility for causing the conditions that led to cannibalism. It was akin to retreating in horror as the bloody man you just beat with a pipe tried to sit next to you in a cafe.

Herndon spent Christmas of 1851 in Tefé, which by then was a hub of organized commerce, with several public buildings and a burgeoning downtown. He was the guest of honor in the home of the Frenchman Jeronymo Fort and was greeted with great ceremony by the region's military commander. After midnight mass on Christmas Eve, a display was held on the lake, with canoes and floating logs lit with candles and lights. Others shot off fireworks, which is still a Christmas tradition in many parts of South America.

Herndon took special note of one particularly attractive woman:

> "The Brazilians, as a general rule, do not like to introduce foreigners to their families, and their wives lead a monotonous and somewhat secluded life. An intelligent and spirited lady friend told me that the customs of her country confined and restrained her more than was agreeable, and said, with a smile, that she would not like to say how much she would be influenced in her choice of a husband by the hope that she would remove to another country, where she might see something, learn something and be somebody."

The jungle can be a lonely place.

We had supper at an outdoor café on the waterfront. It took some convincing to explain to the waiter, through Noella, that I did indeed want exactly what Ian and Jeff ordered, even though they had ordered two different plates. I outweighed the other guys by an easy 50 pounds and, considering the number of calories we burned on a daily basis, I was at a severe deficit. Given the chance to make up some depleted stores, I took full advantage. Noella looked on in amazement as I easily downed both full plates of food, then treated everyone to ice cream. I had another snack or two from a bakery as we walked back to the hotel.

We fell asleep in clean sheets to the sound of a humming air conditioner. The only disappointment was that the internet service was too slow for Ian to video chat with his daughter.

Breakfast was some consolation, as our table was presented with three huge platters piled high with scrambled eggs. I immediately slid one of the oblong plates in front of me and was about to dig in when Jeff interrupted my gastronomic revelry with some tragic news: the plates were serving dishes for all the hotel guests, and we were expected to take only portions reserved for normal human beings.

Goddammit.

I did as I was told, but once the waifs at all the other tables had their paltry fill and made motions to move from the table, I dished all of the remaining eggs onto my plate. After breakfast, we got another full meal at a café and ran errands around town.

Tefé's downtown offered quaint cafés, restaurants, grocery stores and all other trappings of modern life. I found the bakeries particularly attractive and did my best to contribute towards their rent. With more time on our hands, I would have enjoyed touring the old brick colonial buildings.

Once back on the Navy barge we cleaned and organized our gear. Just before we shoved off, a sergeant told us about some recent assaults that occurred downriver, marking the locations on our map while providing disturbing details of each violent act, complete with graphic renderings of severed limbs and gunshot wounds.

Okay, thank you. Time to get going.

Shooting the Amazon

Day 85, Mile 2,711

"He wasn't anti-gun really—he just wasn't pro-being-shot-at."
Adam P. Knave

"Can any of you shoot?"

It had been eight months since Pamela Wells, the executive producer of National Geographic Television, asked me that question in her corner office in Washington, D.C. I crouched on the too-low-too-plush couch, wearing my brand new, excessively stiff linen sports jacket, navy slacks, over-starched dress shirt and not-yet-gravy-stained tie. Wells was asking about the team I'd assembled to run the Amazon from source to sea as I described their attributes with pride.

Now, on a white sand beach on a treeless island in the middle of the Amazon River, 100 miles downstream from Tefé, her question seemed much more relevant. Jeff, Ian and I were gazing down the barrels of three firearms of varying vintages. On display were a sawed-off lever-action rifle, a revolver (somewhere between .22 and .38 caliber) and a single-shot 20-gauge, all in the trembling hands of pimple-faced adolescent boys trying their damnedest to be tough. A fourth kid brandished a machete. The guns weren't particularly high power, yet soft lead hurled at a decent speed from any gun, regardless of its paltry caliber, can cause problems with vital organs and ruin your entire day.

The three of us stood back, as ordered, with hands clear and refraining from sudden movements while the pubescent bandits searched the hatches of our kayaks and demanded the drugs they were most certain we were hauling down the Amazon. Our two shiny white kayaks carried short poles bearing the famous green and blue National Geographic flag—a clear sign of our elite drug-runner status. Exchanging glances, the three of us thought the same thing: please do not find the white protein powder sealed in clear plastic baggies. If they found the powder, we seriously doubted the mood would remain as pleasant. Such a discovery would probably find us in one of those "opportunity for growth" moments most would prefer to avoid.

Pamela Wells had been in a justifiable hurry. She was an important and busy woman. I was rushed into her office by her assistant, passing cubicles with Emmy awards on top of the short, thin movable cubicle walls. I was surprised at how tall the gold statuettes appeared in person and pleased to see the honorees had used high-quality clear packing tape to secure them properly. The headquarters of National Geographic is a working office building with little effort toward aesthetics once you get past the grand foyers and historic exhibits on the first floor. Wells' office was strewn with papers and dry erase boards. She didn't so much *sit* in her desk chair as hover above it, while rushing from one important meeting to another, barely giving her denim jeans time to warm the naugahyde. When her attention turned to me, her singular focus was how to turn an Amazon expedition into good television. She just needed to hear "the catch."

Well before selling what remained of its soul to Fox broadcasting and then becoming a suburb of Disneyland, National Geographic TV specialized in productions involving people doing manual labor or naïve naked couples starving themselves in the outdoors. I wasn't too sure that actually risking our lives doing something that had never been done would garner the same attention as a tow truck driver in Alaska or a whiny naked tattooed yoga instructor in a Florida swamp.

Wells cut to the chase.

"What dangers will you encounter?"

I rattled off a list of feral animals, deadly rapids, weather, narco-traffickers, giant waves, bad coffee, illness and, of course, pirates.

"Can any of you shoot?"

Until that meeting, any discussion of firearms had been kept within the hushed confines of the expedition members. We didn't want to alarm family members and, if we did opt to carry guns, they would most likely have to be obtained illegally. So, sitting there in my crisp, new Jos. A. Banks business casual, I tried to come across as a well-informed and intelligent professional. I didn't want to reveal the latent redneck tendencies I did my best to tamp down in college and various cubicle-constrained careers. A direct question about shooting was the last thing I expected in the staid confines of the National Geographic Society headquarters. Thrown off my game, I scrambled for a white-collar answer.

"Well, we are all from Texas and, of course, we can shoot."

Wells and her assistant took their own half-second pause, glanced at one other, then burst out laughing. Finally, Wells composed herself and said, "I meant can any of you shoot with a video camera?"

Goddammit.

Now, eight months later and a hemisphere away, we stood a good chance of getting shot because we couldn't explain to these punks that the white powder in our hatches is protein drink powder, not the stuff that looks just the same, and for which they were willing to kill.

Can I shoot? Give me a gun and I'll shoot every one of these sonsofbitches.

The gunboys watched us come ashore, then parked their two motorboats about 100 yards up the beach and came running and yelling over a low dune, guns pointed and machetes swinging. We backed away from the kayaks as they stormed forth. There was nowhere to run and nowhere to hide.

After going though most our stuff and throwing it on the sand, they headed back, empty handed, to their boats; then as an afterthought, the little guy who seemed to be the leader came back and grabbed our video camera.

The incident made me wish we'd packed heat. Seeing those guys running toward us with guns, it would have been nice to fire a couple of warning shots to turn them around. A bit rattled, we paddled about a mile to the next village as the sun set and were invited to camp on a moored fishing boat for the night.

Throughout the next day we debated the merits and liabilities of arming ourselves, and what would, should or could have happened during the altercation with the punks. Jeff and I were pretty much of the mindset that anyone running toward us with guns pointed had opened themselves up for target practice. Nothing says, "shoot me!" more clearly than pointing a gun. I figured we could fire a single warning shot, then kill them only if they kept coming. My reasoning was that anyone who had the gall to point a gun at me had better consider the possibility that I might shoot back. I hated the idea that my life was in the hands of some teenage punk, so I would just eliminate the variable. Ian, on the other hand, figured the boys just wanted our stuff, and he wasn't prepared to kill anyone over mere things.

Around late morning the following day, an aluminum jon boat with a good-sized outboard passed us going upstream, then turned around and came right at us. In it were three guys wearing dark baseball caps, colorful surf shorts and bulletproof vests. They carried a lever-action 30-30, a semi-auto pistol and a pump shotgun. None were aimed at us. They appeared to be an ad hoc police patrol or vigilante force. I handed over the yellow tablet with Betina's Portuguese phrases. The clean-cut leader seemed immediately satisfied, then asked if we'd had any trouble. I did my best to explain the night before, which he seemed to take in stride. He also said we shouldn't have any problems from there on down, which, to his credit, we didn't.

At one point as we held onto the sides of their boat while sitting in our kayaks, he turned to talk with Jeff, causing the barrel of the rifle in his lap to point straight at me. I politely asked him to move it. He laughed a little, apologized and shifted the rifle.

In 2016, Andrew Zeiss, Maddy Bohler and another friend were making their way down the Amazon from Tabatinga to Manaus, paddling an unpowered pecky pecky. In Coari, which lies in this section, they rented space in one of the floating huts, built on huge balsa logs that crowded the shoreline. In the middle of the night they were awakened by a dog barking. Looking out the top floor window, they saw a group of armed men approaching in a canoe. The three North Americans hunkered down and did their best to disappear, hoping the boat would pass.

Instead the men charged upstairs and burst screaming into the small room. Maddy and her friend were able to escape by jumping out the top floor window into the river and swimming to shore. Andrew was pinned down by the attackers, kicked and beaten with fists and pistols. His leg was broken and blood poured from several large gashes in his head. They left him lying semi-conscious on the floor and, after rifling through his gear, discussed plans to hold him for ransom. This lit a fire in Andrew, who jumped up and fought them all over again. This time, he was able to stumble out of a window into the river and swim to another floating house, yelling for help the entire way. He beat on several doors, swimming from one house to another and was turned away each time. Finally, a man gave him a ride to shore in a canoe.

Andrew stumbled into town, covered in blood and dragging his broken leg. All the businesses were closed for the night and the streets were empty. Finally, he found an open bar and shot through the door, shocking the small crowd, to say the least, with his blood-streaked face. A patron propped Andrew on the back of his motorcycle and drove him to the hospital, where he was reunited with Maddy and her friend.

Andrew spent two days in the hospital while the police did their best to ignore their pleas for an investigation and justice. The attackers never found their passports and credit cards, which were mixed in with their food packets, though they took Andrew's violin and all the other gear. The hospital wanted to get rid of them but wouldn't front the money for a flight to the hospital in Manaus. Eventually, Andrew came up with the money and was flown to Manaus for proper medical treatment, where his leg was cast and his head pieced back together with 30 rather large stitches.

Maddy and her friend returned home, while Andrew continued traveling around Brazil after recovering at a friend's house in Manaus. He dropped by my place in Austin later that year to share some tacos and show off the Frankenstein scars across his scalp.

One of the more sobering acts of violence took place in 2017, when Emma Tamsin Kelty was murdered a few miles downstream of Coari. The 42-year-old Brit, having hiked the Pacific Coast Trail and skied to the South Pole, was attempting to paddle the Amazon alone. Several experienced Amazon explorers, including myself, warned her in no uncertain terms of the dangers in this specific area, but our warnings fell on deaf ears.

She was camped on an island not far from Coari when a handful of gunboys came ashore in the middle of the night. One of them shot twice into the tent where Kelty was sleeping. When she emerged, wounded and pleading for her life, the gunboys raped her, stabbed her and cut her throat. Then they took her onto the river in their motorboat and dumped her overboard. They were quickly rounded up after inadvertently sending a satellite distress message while trying to make sense of the electronics they'd stolen. Within weeks the ringleader was slain by a rival gang and the others were jailed. Kelty's body was never recovered.

Chapter Twenty-Seven
Kindness and Comfort in Camara

Day 88, Mile 2,926

> "That boy is your company. And if he wants to eat up that tablecloth, you let him, you hear?"
>
> Harper Lee, *To Kill a Mockingbird*

The armed punks left us a bit unnerved, and Ian suggested we stay in a town the next night, so we ramped it up to nine or 10 mph to get to Camara before dark. As our navigation expert, Ian was convinced the small town had a beach. Jeff and I doubted his prediction since no towns on the lower Amazon had beaches. We got to Camara, climbed the 30-foot muddy cliff (see, Ian!) and handed over the book of phrases to the first people we found. A guy read it and shook his head, which we took to mean, "I don't have the authority to comment on this." We asked for the *jefe* and were directed to a beautifully built and decorated wood-frame house on stilts.

The town is named for an herb called camara, which Maw noted as the cruzeiro plant when he passed this way in 1828. Taken as a tea or in a vapor bath, it was said to cure dropsy, a swelling of the legs and pain in the joints now linked to congestive heart failure. Smyth camped here the night of April 30, 1835. He referred to the river joining the Amazon from the south as the Rio Purus, as it appears on modern maps, but also notes the river is referred to as the Rio Camara. He commented on the lack of beaches on this section of the Amazon (see, Ian!) and tied his boat off to a tree in the fast current just below the confluence. This is where we met Jo and Francisco some 177 years later.

As we approached the house, it was clear this wasn't like any other place we encountered, though truth be told we hadn't yet interacted with many Brazilians due to our pace. The front room was as wide as the house with neat shelves stocked with normal *tienda* fare: canned meats, soda, cleaning supplies. It was neat and clean without a speck of dust or dirt until Jeff and I wandered through the door.

Jo and her team review the maps with West. Photo by Jeff Wueste

A woman came from the back room with a huge smile on her face, the first of many incredibly pleasant surprises. She read the phrases in our little yellow book and her smile grew wider. This was Jocelia Sigueiro, but she introduced herself as Jo and immediately invited us to stay in her home. Her husband, Francisco Poina, came from the back room, exchanged some words with Jo and nodded his agreement. Then they wanted to see the boats, which we planned to tie off to a tree as Smyth had done all those years ago. Nothing doing. Jo directed some of the younger men down the bluff to help us up with our boats. These boys weren't all that big but they sure were strong. With the three of us barely helping at all, they hauled those fully loaded boats up to Jo and Francisco's lawn in minutes, then discussed where to pitch the tents. Throughout all of this, Jo insisted we stay inside the house. We declined since we smelled so bad.

It was clear that Jo was in charge and was doing a great job of running whatever business they had going. Their compound, we discovered, was just upriver of the town of Camara. The evening we showed them our journey on our maps and they were thrilled to see that even tiny Camara was marked on the big Brazilian Navy map. Jo served cold orange soda—pretty heavenly—and the translation book made the rounds as some of the younger guys hung about to hear the stories.

West and our hosts, Francisco and Jo with the translation book, in Camara. Photo by Jeff Wueste

Francisco and Jo made their living harvesting açaí berries, squeezing them and selling the concentrated juice in Manaus. Jo showed me a video of the process on her smart phone, along with photos of their houses during high water, which were astounding. The Amazon comes up to their third step and stays there for three months, which was a 40-foot rise across a river 10 miles wide.

Later, Francisco took us to the next house to shower. The trek involved a wooden catwalk about six feet off the ground. The planks were thin and springy under my temporarily svelte 190 lbs. The shower was a bare faucet on a workbench outside the house. Francisco filled up a 10-gallon tub with naturally warm water, and then dumped in a small bucket of ice. We then used the small bucket to ladle the ice water over our heads, then to rinse off the soap. We all agreed this ranked as one of the best showers of our entire lives. The cold water against our dirty, hot skin was truly remarkable.

At the back of the immaculate house was a large kitchen with a dining table. It would have fit perfectly in an American farmhouse at the turn of the last century: clapboard walls, wooden floors and rough-hewn countertops. The house was built to last. At the table, Jo served rice, canned meat, eggs and grape soda. It was one of the best meals of my life. We ate it all, and when she heard us stacking dishes, Jo rushed in from the living room where everyone was watching Brazil take on Chile in *fútbol*. Ever the consummate host, she shooed us into the living room, insisting we relax. It was all quite surreal and pleasant. After the dishes were cleaned and stacked, Jo showed me photos of her 24-year-old daughter vacationing in Florida and New York. Quite the renaissance woman, Jo was reading Steve Job's biography in Portuguese, which was still on the *New York Times* bestseller list back home. Our world is far smaller than it appears at first glance.

The night was absolutely beautiful, with our tents lined up on the elevated wooden porch overlooking the Amazon under a bright moon. Lightning, as usual, flashed in the distance and fireflies danced around our tents and throughout the looming jungle. A rare cool wind provided inordinately comfortable sleeping conditions until about 3:45 a.m., when Francisco woke us with warning of an impending storm. We dragged our screen tents into the front room, tried to sleep another hour against the thunder and lightning, then gave up. Bleary-eyed, we sipped homemade hot chocolate and coffee while waiting for the storm to abate. Jo and Francisco's hospitality renewed our energy and spirits tenfold.

We got away from Camara at about quarter to eight, quite late by our standards, and enjoyed an uneventful day on the water, which was very welcome. Our camp, which we chose rather late, was on a low-lying beach—an uncomfortable spot given the Amazon's propensity to suddenly rise and fall. So, while Jeff and Ian set up the tents in the dark, I wandered up to some houses to ask permission to trespass. I made my way through two or three acres of hand-tilled crops: corn, melons, gourds, beans and what appeared to be tomatoes. A standard farm dog barked impotently as I shuffled from house to house to find anyone who could grant us permission to stay. Being bone-tired really helped when coming into contact with dogs and others who may mean harm, as I really didn't give a damn. The permission inquiry was a mere courtesy.

Finally, I came to a lit-up house, where a young boy opened the door for an old woman to nod her permission. I figured my Portuguese was getting a bit better, since I repeated the same phrase each evening, *"Podemos acampar aqui por uma noite por favor?"* (May we camp here for one night please?)

Then, regardless of the answer I received, I assured them, *"Nos saíremos de manhã do nascer do sol."* (We'll be gone at sunrise.)

A small group of boys and men followed me back to the camp, but I paid them no mind having obtained the old woman's permission. Old women everywhere garner a great deal respect and authority. If challenged I would simply have mentioned the old woman in order to notify the others in the universal man club that a woman has spoken and I'm here on her authority, so they had best back the hell down, if they know what's good for them.

No one asked questions, which was fortunate for all involved, due to my mood. The jungle rot crept back onto my feet and a painfully abscessed bedsore on my butt worked through every layer of skin I held in reserve. With Ian doing most of the water filtering, Jeff drew the short straw and got to patch the oozing sore with gauze, held in place by industrial strength duct tape each morning. The ad hoc first aid proved entertaining to Jeff and Ian each evening when it came time for me to remove the powerful adhesive patch with one swift eye-watering, scrotum-shrinking move—the result of which was my own personal version of a Brazilian wax. I was peeling duct tape residue off my hairless right cheek for weeks after the expedition.

Manaus

Day 90, Mile 3,077

> *"January 5, At 3 a.m. we passed a rock in the stream called Calderon, or Big Pot, from the bubbling and boiling of the water over it when the river is full. We could hear the rush of the water against it, but could not see it on account of the darkness of the night."*
>
> William Herndon, January 5, 1852

After a very pleasant and uneventful two days on the river, we uncoiled at 4 a.m. for the push from Manacapuru to Manaus, a mere 60 miles downriver. As the river widened, the feeling of progress diminished since we were too far from the shoreline to naturally gauge our speed. Our navigation skills were vastly improved and we always sought the fastest flow, while staying clear of the shipping lanes. Fast boats and general river traffic increased exponentially, indicating our proximity to Manaus. Twice, a passenger jet boat buzzed us from the rear, startling the bejesus out of us as it veered away at the last minute, sending its wake crashing over us from behind, to the jerkish delight of its pilot.

Manaus is a city of 1.8 million built on the northern shore of the Rio Negro near its confluence with the Solimões, which lies about two miles south of the city center. To avoid a difficult upstream paddle, we approached through a narrow cut that bypassed the last few miles of the Solimões and deposited us into the Rio Negro, directly across from the city.

From the river the cut looked too shallow. We almost floated by it but made a last-minute call to ferry across and check it out. The maneuver had us crabbing sideways for a quarter mile across the current as a massive ocean-going freighter bore down with alarming speed. The Solimões was hauling ass around the outside of a massive bend, when we decided to try for it. Making a quick U-turn to face upstream, we went balls to the wall to hold our position against the current while angling slightly toward shore. In this way, we side-slipped across the river without losing ground. It was a lot of fun and nice to use our skills after 3,000 miles of straight-ahead paddling.

In the cut we immediately hit dead water. The low water meant we had the channel to ourselves, save for the smallest of jon boats. We passed through low terrain reminiscent of the salt grass meadows along the Texas coast, and 45 easy minutes later we were in the Rio Negro.

While the name of the Amazon hails from lofty reports of mythical warrior women, the name Rio Negro is purely descriptive. The water is black and shines with a strange iridescence. Gradually, the noises and smells of Manaus reached our senses as we ferried across the expanse of dark water toward a waterfront bristling with cranes and bathed in late afternoon light. Diesel vapors from Godzilla ships moored high in the river wafted over us, surfacing memories of bus stations and construction sites. The screech of steel against steel carried farthest across the water, but as we drew closer even that sound was lost in the general hum and drone of an ultra-urban beehive driven by the constant dirge of internal combustion. Paddle strokes, once the only sound for days on end, faded to insignificance.

The cut from the Solomoes into the Negro. Photo by Jeff Wueste

We slipped behind a tug pulling a barge and crawled up the eddy along the opposite bank, stopping at a succession of floating gas station convenience stores to ask directions. Finally, we pulled out of the kamikaze harbor traffic to call Barbara. Our landing spot was a scrap of muddy beach at the base of the massive concrete sea wall built to prevent the river from sweeping the metropolis away. It proved a perfect gathering spot for a landfill's worth of household garbage. Each passing boat sent waves of plastic, wood and styrofoam swirling around our feet. Stepping gingerly, I punched my sister's number into the sat phone. Within minutes, Barbara, who knew more about our position from a continent away than we knew ourselves, reported that we were close to the naval station. It was just downriver of the brand-new Rio Negro suspension bridge which was a landmark we couldn't miss, decorated as it was with artsy lights and looming delicately against the evening sky.

On the navy barge, Lt. Leita confirmed our legitimacy with a quick cell phone call to Cmdr. Fernandes, then invited us to store our kayaks on the barge, which was tethered to the shore by 60 feet of cable. We all clambered into a jon boat and pulled ourselves along the cable to a flimsy wooden dock, breaking a passage through the trash like Amundsen through the ice, if ice smelled like shit. On the dock, a handful of enlisted men and smartly dressed officers engaged in a long, unsmiling conversation, which included quite a bit of gesturing in our direction. Finally, they informed us of the problem: it was our shorts. They weren't allowed on the base. I offered to disrobe immediately but was stopped by all involved as soon as they realized I was fairly serious. Eventually, the head cheese barked orders into a phone, and moments later a truck arrived.

We settled in for the ride to our hotel across town, as the truck drove slowly through the beautifully landscaped naval base toward the ornate front gate 50 yards away. The truck stopped. Salutes were exchanged and we were ordered out of the truck. We quickly realized the truck ride was merely so our bare legs wouldn't be seen crossing the cobblestone parking lot from the dock to the front gate. A small taxi just outside the gate was our ride to the hotel. It was our own personal river-scum ride of shame. We were the crazy aunt put away in the attic, the drunken brother-in-law at the Thanksgiving table: not bad enough to kick the hell out, but not good enough to let the neighbors see. Perhaps we should've shaved our legs.

We drove forever through Manaus in the air-conditioned taxi to our hotel, Chez La Rois, where Jackson and Viviana met us at the gate and showed us to our rooms. Jeff's girlfriend, Sheila wasn't due to land until 11:20 p.m., so we showered and went to eat a huge amount of food at "Burguers and Beer." The beer was cold, the burguers excellent and the attractive patrons, oblivious to our presence, were a nice distraction. The modernity and size of Manaus caused our heads to spin, but we looked forward to a couple of days rest.

William Smyth landed in "Manoas" on May 3, 1835:

> "At ten o'clock we arrived at the town of Barra, which is now called Manoas. It contains about one thousand inhabitants, nearly all Indians; but the population is said to have been on the decrease ever since the mutiny in 1832, when the Commandant was killed, and the town has the appearance of being on the decline, for many of the houses are in a state of decay."

During his extended stay in Manaus, Herndon, always with an eye out for the fairer sex, was quite taken with Donna Leocadia, *"the pretty, clever, and amiable wife"* of Enrique Antonii. The Italian-born Antonii had lived for years in Brazil and was a great help to Herndon. Donna was an intelligent and forthright woman who sat equally with the men at the table and participated in the conversation, which was unusual in that time and place. She gave birth to a son during Herndon's stay and asked him to be the godfather. However, the church *would not give its sanction to the assumption by a heretic of the duties belonging to such a position."*

When Maw visited 25 years earlier he took note of a three-year-old little girl who sat at the head of the table during a formal dinner hosted by the regional high-court judge. It's not clear whether this same precocious girl was Donna Leocadia, but forthrightness rarely departs, once it is established. Maw, by his own account, rarely came upon "women of class" in Manaus or along the route.

In 1828, the tradition of keeping wives and daughters out of the public eye extended even to the dinner table when formal guests were present. As for women of ordinary circumstance, Maw held a low opinion of their appearance and sexual proclivities, as evidenced when he encountered skinny dippers:

> "Going down a good look-out was always kept on the padre's glass, and if it happened to be brought out whilst they were in the water, a rush was immediately made to get further in, or to run out and hide themselves. This was pretty nearly the amount of these females' modesty – chastity not being a virtue for which they are at present celebrated, or are likely to be celebrated during the continuance of the now existing system. Their alarms at the glass were however generally causeless; for, laying aside its not possessing the powers they attributed to it, they were not sufficiently fascinating to occasion the indecorum they imputed to the padre and his friends."

Lapping up Luxury

> *"Fashionably amusing table manners are a matter of breaking the right rule at the right time."*
>
> P.J. O'Rourke

We slept until about 8:30 a.m., which would have found us on an average day at least 10 miles downriver. During breakfast in the pleasant hotel patio, I clumsily handled the tiny china plates, delicate glasses and silverware with my swollen, cut hands. My sausage-like fingers wouldn't fit through the handles of the coffee cups and I was afraid to break something every time I rose to retrieve more food from the tiny buffet or reach across the table for another petite something or other. Someone on the team (not naming names) loudly passed gas in midsentence without batting an eye. A bit of refinement and readjustment to polite society was in order.

We took the requisite tour of the famous opera house downtown. During their visit in 1986 Joe Kane wrote that the formerly ornate structure was in desperate disrepair. Mozans in 1908 gushed about the grand appearance of the newly built opera house with its ornate architecture rivaling the finest in Europe, though he questioned why such a grand edifice would be built in *"a territory so sparsely populated, and where apparently, there is but little demand for it."* The answer of course was the great piles of rubber money flowing into the Amazon.

The opera house is now fully restored to its original grandeur and updated with modern plumbing and electricity. It is active year-round, with festivals, theater and music. The 2010 census counted 1.8 million people in Manaus; at the time of Mozan's visit it was just under 50,000 and filthy with money, its port bustling with ships carrying rubber to the world and returning with holds full of opulent baubles.

The historic opera house in Manaus. Photo by Jeff Wueste

Mozans remarked about something that I noticed, as well. The lumber used in the crates loaded with raw rubber cargo was pine imported from the United States. This was also true for the lumber for houses and other structures, despite Manaus being smack dab in the middle of the densest forest on the planet. In the Amazon, many different species of tree grow right next to one another, and not all are good for lumber. To pick and choose the right tree, fell it, then build a road to get it to the river or to a sawmill is a huge task. Meanwhile, lumber companies in North America approach timber in the same light as farming wheat or corn. A crop grown and harvested in one place and hauled en masse costs less to produce and is therefore more profitable. Even at the turn of the 20th century, it was cheaper to cut and mill timber in North America and ship it to the heart of the Amazon than to harvest it just a mile away.

The team with Lt. Kenned. Photo by Sheila Reiter

Manaus is now a hub of manufacturing, shipping products throughout the world. To achieve this status, in 1967, the Brazilian government established Manaus as a free trade zone, with businesses able to establish bases that can operate without paying taxes. The *Zona Franca de Manaus* covers several large municipalities in northern Brazil, with the goal of attracting large business and industries to the area. The effort has been quite successful, judging by the cosmopolitan look of Manaus. Here in the center of the rainforest, over a thousand miles from the Atlantic Ocean, supertankers and mind-bogglingly huge container ships moor in the harbor waiting for loading and unloading. Harley-Davidson builds motorcycles here and ships them around the world. Shops and restaurants found in Denver, Hong Kong and Paris swarmed our senses the day after we ate freeze dried food from an aluminum bag on a muddy bank just a few miles upriver.

An additional bonus is the weather. While it's a common belief that the jungle is a steamy hotbed of misery, we verified what most other explorers came to report: that the weather here is much nicer than most places in North America. The climate is comfortable year-round, with short-lived storms bringing relief from the heat, without the devastation of coastal hurricanes.

In Manaus we also enjoyed the company of Jeff's hunka-hunka-burnin-love, Sheila, who brought a softer touch of civility to the expedition. She brought care packages complete with the necessities: M&M's, Cheetos, beef jerky, nuts, computer hard drives, a replacement video camera, nice notes in cards and pictures of Ian's incredibly cute daughter, Addie.

During our three-night stay in Manaus, we came to the harsh realization that the once parenthetical world of our expedition had quietly turned the tables on our pre-journey lives. Our new normal included daily gastrointestinal issues which we now gave little attention, yet the same symptoms sent others into states of despair. We pitied those who complained of loose bowels, when, within mere steps they had the luxury of a toilet with ample rolls of toilet paper. What heaven! All ambient noises came at us much louder, causing us to jump now and then or at the very least jerk our heads in the direction of sounds that went unnoticed by those unaccustomed to the natural silence that filled so many of our days. Where once, thousands of miles and millions of minutes long past, we were inundated with the verdant smells of greenery and life that hit us like a organic wall after descending the Andes, we now struggled to cope with the myriad odors emanating from kitchens, garbage cans, internal combustion engines, perfumes, industry and who knows what else in this urban jungle through which we now sauntered, wide-eyed, like boys just off the farm for the first time.

The ultra-modern, multi-storied, air-conditioned shopping mall sparkled with masses of Apple and Dell computers, diamonds, furs (yes, fur coats in the Amazon jungle), high fashions from around the world and the latest pieces of electronic gadgetry and knickknacks that no sane person could possibly live one more minute without. The cheeseburgers were plentiful, as was the hot water from the shower spigot. We were easily lured in by these lavish sirens, yet with every amazing meal or day waking in clean sheets, the urge to escape grew stronger, leading us into fits of irritableness or forced silence. It was time to get back to our life on the river.

The super modern Manaus shopping mall caused a shock to our dirty souls. Photo by West Hansen

Chapter Twenty-Eight
Monsters on the Amazon

Day 93, Mile 3,084

"Whoever fights monsters should see to it that in the process he does not become a monster. And if you gaze long enough into an abyss, the abyss will gaze back into you."

Friedrich Nietzsche

Before we left Manaus the owner of the hotel Chez Le Rois, Mauro, took us to run some errands. While driving us through the city he spoke of adventures shared with his father up and down the local rivers. His father captained the river all his life, having worked with Jacques Cousteau and other notable explorers. Mauro told us about the time his father took him to see a waterfall on the Apuau River, where they spotted what looked to be a tractor tire on the riverbank. On closer inspection the tire turned out to be a massive anaconda, coiled and resting. When they came within two meters (*two meters!*) the snake disappeared into a large hole, quick as a flash. Mauro was surprised by the speed of such a huge snake and from then on gave it a bit more respect. He educated us about the Rio Negro, including the fact that it contains the largest freshwater archipelago in the world and a huge underwater canyon just upriver of Manaus featured in the documentary, *Amazon Abyss*.

He may have been referring to the Hamza River, a subterranean body of water flowing about 13,000 feet under the surface of the Amazon River. This huge underground river flows only about 30 feet per year and is far larger than the Amazon, extending about 3,700 miles, roughly tracing the Amazon's path. The Hamza is a fairly new discovery, so not much is known about it. It is supposed to spill out into the Atlantic, and I wondered what a submarine capable of exploring the waterway would look like.

Mauro's father guided Martin Strel's team down the river when the Slovenian adventurer swam some 3,273 miles from Atalaya to Belém in 2007. Mauro's old man was with the boat crew that looked after the swimmer, keeping buckets of fresh blood handy to distract hungry piranha should the need arise (it didn't). Through his connections, Mauro learned that the area around Garupa, downriver from Santarém, was teeming with river pirates. He made a call to introduce us to the guy who provided Strel's security and to see what it would cost for him to escort us through the volatile region. The man wouldn't name his price up front, and having paid several thousand dollars to dance the Peruvian Shuffle already, I told him if he couldn't give us a firm cost then we'd just take our chances with the pirates.

We prepped the boats all afternoon, then paddled past the city as the sun set over the jungle. Though it was completely dark, the lights from Manaus provided ample light to paddle almost to the mouth of the Rio Negro before pitching camp. Our plan was to sleep a little later in the morning, in order to pass in full daylight, the famous twain where the latte-colored Solimões meets the Rio Negro's black waters. Here the great river again becomes the Amazon, after a 1,000-mile purgatory imposed by Brazilian cartographers, who call the section from the Peruvian border to Manaus the Rio Solimões. On Peruvian maps, the Río Amazonas begins far upstream at the confluence of the Ucayali and Marañón.

It rained hard most of the night and we felt silly hunkering in the mud across from the luxury of Manaus. We could make a call and be in a hotel via a water taxi within 30 minutes, but it would mean more time to get moving in the morning. Besides, none of us felt like raising our profile. We were all a bit unnerved by Mauro's tales of the danger waiting downriver and I suggested we get guns in Santarém if we couldn't afford to hire an escort boat. No one objected, though we all agreed that an escort was the better option.

Jules Verne's *800 Leagues on the Amazon* makes many plot twists in Manaus near the very spot where we camped. It was here in the book that a diver searching the river bottom for a dead body got into a tussle with a giant electric eel. The attack came just as the diver passed the skeleton of a 20-foot caiman within sight of the corpse he was seeking. Though a strong proponent of science, Verne never let facts get in the way of a good story.

The River Amazon

Day 94, Mile 3,197

> *"I always felt some anxiety in crossing so large an expanse of water, where violent wind-storms are frequent occurrences, in such a boat as ours. Our men, with their light paddles, could not keep such a haystack as our clumsy, heavy boat either head to the wind or before it, and she would, therefore, lie broadside to in the trough of the sea, rolling fearfully and threatening to swamp."*
>
> Lewis Herndon, 1852

Herndon crossing the mouth of the Madeira from "Exploration of the Valley of the Amazon" by Herndon and Gibbons

In spite of our best intentions, we missed actually seeing the famous confluence of the Rio Negro and the Solimões, as it was still dark and overcast when we departed. We simply couldn't wait around until daybreak to get started, after being sedentary for three days. Once underway, we paddled 68 miles without once getting out of the boats, and camped opposite Mauari, about 10 miles shy of the Madeira River. The next morning, we woke at 3 a.m. and pushed through another hard-earned 68 miles.

The most difficult task for the day came as soon as we launched into the darkness. Not only was the dark expanse of swift moving water several miles wide, but we had to cross a busy shipping channel to get to the first cut, just upriver of the Madeira confluence. All of our night-paddling experience had been in tight swamps or small river settings, so the enormous width of the Amazon, where it spread out at the mouth of the Madeira, combined with the threat of some giant dark ocean liner silently creeping into our world, had us on edge. Ships ran without lights on the Amazon, since their GPS, radar and sonar offer far better navigation than any beam of light. There was no real advantage to going faster, as my GPS indicated we were already moving at a substantial eight miles per hour, but we didn't tarry in our push to get to that cut on the opposite bank, through which we hoped ships would not be able to travel.

Gradually, we were able to make out more detail along the rocky banks and tall white cliffs on the river's north bank. While we were still unsure of our distance from the shore, this was at least some indication that we were substantially closer. We didn't want to be too near, as there may have been shallows or rocky outcroppings, but we did want to be close enough to shore to be certain we were out of the shipping lane.

Herndon crossed the Madeira-Amazon confluence on February 19, 1852. Before reaching the Madeira, he floated past a checkpoint as officials on the bank shouted their demands to see his papers and search his small boat. Unable to paddle upstream against the current, he simply drifted by. The officials chased him down to examine his papers, which they found to be in order. Given the inane bureaucracy and his own surly crew, he wasn't in much of a mood when authorities later summoned him back across the Madeira for another bout of paperwork.

Herndon had taken to traveling at night, when the weather was calmer, rather than fight the strong head winds and large waves that filled their afternoons. However, given the river's ocean-like breadth he opted to wait until first light to make this particular crossing. Despite decades of experience sailing the high seas, he was wary of the big river.

The Franciscan Brothers gave the Madeira its name in 1636, a rather pragmatic moniker based on their observation that the river disgorged large mats of logs into the Amazon. It flows for 2,000 miles and carries 15 percent of all the water in the Amazon basin. If it flowed into the ocean rather than the Amazon, it would be counted as one of the 10 largest and longest rivers on the planet. While the Portuguese called it the Madeira, Spanish explorers referred to it by its indigenous name, Cuyari, on maps published in the 1700's.

We had come about 3,280 miles from the source in 95 days. While paddling bleary-eyed through the pre-dawn darkness, I realized how much our daily world had changed in that time. An awful lot had been crammed in since the freezing weather, dry hacking coughs and altitude sickness in the Andes. The finish was too far off for us to even consider as we settled into the Zen of our well-oiled daily routine.

Each morning we saddled up and eased into our predictable watery treadmill. The first minute or two we drifted, the boys waiting for me to give notice that the odometer on our last functioning GPS unit was reset and ready to tick off the miles, like a bucket drained one teaspoon at a time, then refilled the next morning. Long gone was the time when I monitored our progress on the digital face every few minutes. To do so was maddening, either in the discouragement in our slow pace or false hope when we hit a sudden, yet all too brief, swift patch. None of us considered the beginning or end of the journey. I only checked the digits whenever we stopped for lunch, stopped for the day or changed the batteries when given the warning. Mind you, the treadmill wasn't disconcerting. It was just our life and we all kept the bigger picture close to our hearts, as we discussed mileage, timelines and navigation around supper. We simply ate, slept, drank, talked and paddled.

Sometimes none of us uttered a word for hours, lost in our own thoughts, then someone would begin speaking in mid-conversation on a subject from hours or days beforehand. It was especially exciting if the break in the silence was to produce the answer for a bit of trivia on which we all chewed for the last few days.

We paddled 15 miles before dawn, emerging from the half-mile wide cut into the Amazon proper amidst a flurry of activity from a pod of dolphins playing in the twain below the island. For an hour, we paddled through a blinding thunderstorm, which was somewhat pleasant, save for the terrorizing thought of being run over by a supertanker barreling out of the vast grayed-out horizon. At Itacoatiara, a really large ship moored at the port and gave us a real appreciation of how fast the river flowed as we swept past the enormous steel hull, only a few yards away. Usually, the bank was too far away to get a feeling for the speed of the river.

Maw noted Itacoatiara in his journal as the town of Villa Nova, from which he reported the major export was guarana, a stimulant prepared using a method the inhabitants of the region held secret in order to secure a strong market. Since then, guarana has been used in soft drinks and supplements to give people a kick similar to caffeine. My favorite Brazilian soft drink was *Antarctica* soda, which advertised a good amount of guarana along with cherry flavoring and that ever-important ingredient, sugar. Maw accurately reported that guarana acted as a diuretic and, when too much is ingested, produces "nervous irritability."

More often than not on these days, the Amazon handed us our asses. It was only through our steady incessant zombie paddle strokes that we made progress. Each of us was afraid to take more than a few strokes break, to grab a bite or drink, for fear of becoming far too comfortable in the rest. "Beware the chair" is the mantra of the ultra-runner. Discomfort is best endured if there's more paddling to do and it made the days' end far sweeter.

The bow deck of the author's kayak and his viewpoint for 3,800 miles. Photo by West Hansen

The equatorial fever abruptly broke towards the end of each day as the waning sun dropped beyond the endless flat green of the jungle. Temperatures plummeted well down into the 80s, which felt amazing against our fried skin. Thousands of parrots and other chatty birds gathered in the canopy, floridly gossiping about their days' adventures in the short time they had before the night turned them from predator to prey.

Evening waves of cicada chatter showered over us from the jungle's edge. Known as "katydids" in Texas, the choir produced an immense low roar that wasn't altogether unpleasant. Once the temperatures cooled at night, the switch was flipped and the drone came to an abrupt startling halt, making their presence even more apparent.

Macaw sightings were one of our little daily bonuses. In pairs, they flew in one direction in the morning and back in the evening. This must have been the pattern for all macaws, since we'd be 50 miles downriver when we spotted the macaws flying back. Given the distance, it seemed unlikely they were the same birds. We never spotted pairs of macaws flying in opposing directions. Many birds of the Amazon pair up for life leading to both joyful and sad anthropomorphic effects. Henry Maw, in 1828, noted the pairing of parrots he took from Tabatinga back to England. They are very social little guys, who learned words and phrases, often trying to make friends with mirror images of themselves. While the two were inseparable, he didn't notice how much they meant to one another until he forced the separation of a couple upon his return to England, giving one to a museum and the other to a garden in the park. Within days they both died, though in no ill health.

At the end of these long days, which began around 3 a.m., I was plenty tired and irritable. By now, we needed about two hours just to pack up and get on the river, which took on the characteristics of an endless lake. The current was now barely perceptible at around one mile per hour if we merely drifted. Though we knew it would happen eventually, none of us was prepared for the effects of the Atlantic tides to be felt so strongly almost 1,000 miles upriver of the mouth. It was frustrating to bust ass all day long, with only one 30-minute break for lunch, and make so little progress.

Sitting in my tent four days out of Manaus, I heard something crash in the woods next to our campsite. Perhaps it was just some limbs falling. Normally I'd find such noises unnerving, but at the moment everything was just irritating. We couldn't find a decent beach for camping, so we ended up on a muddy slope swarming with mosquitoes and flying bugs.

It was another low point in my race towards de-evolution: sitting naked in my tent typing on my tiny computer, eating a bag of dehydrated chili mixed with canned meat, inches from a pile of foot powder, swatting bugs and sitting on a towel to dry my butt so my wound could heal. The only relief from this misery was hearing my daughter talk about her life events when I called in for my daily report. We were all drained and ready to be done with this thing.

The Obidos Narrows

Day 96, Mile 3,406

"Thought is Energy."
Louise Tate Schlegel

We never quite got comfortable with the wakes of large ships crashing against the shoreline, though it happened with regularity at every camp spot. More often than not, the sound of breaking waves would wake one of us long enough to shine a light at our kayaks to make sure they weren't drifting away with the retreating surf—which happened twice over the past couple of weeks, though they sat high and dry on the bank each evening. This time the boats didn't get away, but I had to get up in the night to drag my boat farther up the bank. We were up again at 3 a.m., breaking camp as howler monkeys serenaded us from their hidden haunts in the forest. Their huffing, lion-like calls were extraordinary, but if we hadn't known the source of the cries the scene would have been pretty creepy.

Later that morning, as we passed the town of Parintins on river right, a large ship hauling liquid propane gas overtook us, throwing up a colossal wake. We sized up the waves as they approached and just before they reached us we put on a burst of speed. Marathon canoeists surf each other's wakes all the time to gain a little boost, much like cyclists drafting in a peloton. The same technique works with the wakes of big cargo ships; the difference is simply a matter of scale. The waves streaming off that gas-hauler were cresting well over our heads. To catch them we needed to sprint just as the wave overtook us from behind, while taking care not to sink the bow of our kayaks in the trough or get turned sideways. When you get it right the effect is pure magic. The kayak takes off like a rocket, and you surf along, with little effort, at 12 or 15 miles per hour. It's not without a bit of apprehension—you need to stay on just the right part of the wave to keep tragedy at bay—but it sure beats swinging a paddle in the Amazonian heat.

Unfortunately, the wake-riding session lasted only half a mile, and then it was back to no current, steep chop and a stiff headwind. Our speed dropped to less than five mph at times, despite our best efforts. By the end of the day we made less than 60 miles, our lowest daily total since before Iquitos. We were all pretty let down, but a bath (with soap) in the river helped, and we had a nice meeting with the owner of the beach where we camped. Paul (pronounced Powell) came riding down on his horse bareback when he saw us setting up camp. We shook hands, smiled and gave him the phrases out of the notebook. He seemed to get a kick out of it and we took some photos with him and his horse. It was a pleasant evening, despite Jeff learning via a surprisingly strong cell phone signal, that his credit card had been hacked in Manaus, accumulating $6,000 worth of fraudulent charges. There was no news from Mauro about the cost of security through the pirate country, which was drawing nearer every day.

One night along this part of the river, I woke just after midnight to stretch and mark territory. Naked, I strolled to the river's edge in what had to be the most perfect temperature in the world. The moon, not quite full, lit up the Amazon to the point of squinting. For the moment, my aches and injuries remained at a manageable level and I felt pretty good. Being completely dry was such a luxury, albeit momentarily. Looking down, I was surprised to find my middle-age paunch no longer existed. In fact, my love handles had been reduced to almost nothing and the doughy consistency of my midriff had been replaced with what could have been construed, with a bit of imagination, as muscles. Feeling all around my torso and shoulders, it occurred to me that I was actually in decent shape and felt not too bad.

As I took in the midnight scenery, which was absent any noise and gracefully free of mosquitoes, I thought of the historic voyagers who came before me, passing this very spot. Perhaps they experienced the same magic in the moonlight, beheld the same wondrous calm and experienced the same connection with this great river and everything it nourishes. My good fortune, in spite of all the struggle and pain, was not lost on me. I was fortunate to share this unique experience with such amazing figures.

Under the cosmos and the spotlight moon, alongside my sleeping friends beside the river down which traveled ghosts of famous explorers, I felt at once connected with the universe and my forebears, and vastly insignificant.

Juruti

Day 98, Mile 3,406

"The flame that burns twice as bright burns half as long."
                                                    Lao Tzu, Te Tao Ching

As the Amazon widened and narrowed, we did our best to find the shortest course around bends and through the numerous side-channels. We passed inside the island of Cacao, the size of a large town, with about two miles of water separating us from the city of Juruti, which boasted several tall buildings and a considerable waterfront with industry extending out of town.

Toward the end of the day we met a young man sailing a tiny wooden boat upriver. This was Patrick Falterman, who hailed from a small town northeast of Houston, not far from where I was raised. It was a nice surprise to meet a fellow Texan in these parts. Patrick, in his early twenties, was living the life of an adventurer and traveler. In a makeshift plank boat with a plastic tarp for a sail, he was sailing up the Amazon to Tefé, with plans to continue into Colombia and return to Brazil via the Rio Negro. He stopped in towns along the way to work for a few weeks, earn some money, and then continue on. This was Patrick's fifth year in Brazil and he told us the Amazon narrowed at Santarém in a place known as the "throat of the Amazon." This was the most constricted section of the river and we could expect the current to speed up considerably, which was just what we needed.

Though his appearance was straight out of the Dirtbag Quarterly fall collection, Patrick's goals were those of the truly noble traveler. Patrick's parents, whom I met at their dinner table years later, told me he'd struggled with a short attention span in spite of his obvious intelligence. When he announced his plans to travel Central and South America after high school, they let him go his own way. He asked for nothing and accepted no financial help from his parents during his foray, though his mother did send a copy of his passport to the U.S. consulate in Mexico after Patrick had been robbed of all his possessions.

We rafted up with his boat and spoke in our comfortable drawls for a few minutes, though Patrick had adopted a distinctly Brazilian twang to his voice after so much time in the country. Though his gear was quite sparse, he was very self-sufficient and refused our offers of food and coffee. He also refused our offer to camp with us for the night, wanting to make it to Juruti by the day's end. Patrick was an impressive young man, with far more sense of purpose than most his own age.

Texans on the Amazon. Patrick and West. Photo by Ian Rolls

After a few years gathering worldly knowledge, Patrick returned home a better man. His father, Pat, a commercial pilot, helped him obtain his flying license with a goal to return to Brazil as a crop duster, having met a young woman who was the only female crop duster in the country. I came to know Patrick's sister, Ellen, an actress and flight instructor, and his mother, Cindy, a semi-retired special ed teacher. Patrick took my daughter for a ride in his two-seater airplane over the dense piney woods around their east Texas home. Several months after this ride, while flying over the Trinity River with his friend Zach, the engine on Patrick's plane gave out during a steep, low-altitude climb. The tiny plane stalled and immediately nosedived into the deep river, killing both young men as their families watched in horror from the shore.

On the Island of Swimsuit Models

Day 99, Mile 3,462

> *"Women, men, children…spend the greater part of their time stretched in their hammocks in the most complete immobility."*
> Gaetano Osculati, Obidos, early 1800s

The Amazon makes an abrupt change after Obidos, which rests high on the cliffs along river left. As we took the sharp right-hand curve from Obidos, we faced straight into heavy wind and waves. At one point, Jeff noticed I was paddling much slower than normal, though I was obviously working really hard. We pulled over to find my entire back hatch full of water. I hadn't screwed the hatch cover down tight enough, and the pounding waves had done the rest. We drained the water and spent a valuable hour partially drying my sleeping bag and gear in the sun and wind.

Ian picked up my spirits by presenting me the decaying head of a huge catfish with some serious horns coming out of each side. The head was about the size of a large dinner plate and thick as a loaf of bread. The flesh that still remained was a bit odiferous, but not enough to prevent me from mounting it on my bow. Ian soon came to regret the devil fish head, because it stunk up his airspace as I paddled near their tandem during discussions, my bow directly adjacent to his cockpit.

We camped on a small bluff in the middle of a small cluster of ranch houses about 10 miles below the city of Obidos, on river right. A bit desperate to find a place to stop, as I was worn out and dehydrated from fighting the heavy wind chop, we went against our habit of camping in solitude. Jeff thought the far bank looked like it had better beaches and I thought so too, but the other side of the river was five miles away and I was really ready to get off the river for the day.

Lacking a proper beach, we pulled up to a docked riverboat full of some of the best-looking people on the planet. About eight or 10 guys and girls in their twenties were on the boat passing a bottle. A partially clothed woman, poured into a hammock and exuding the languid sensuality of the tropics, regarded us with lazy eyes and a subtle smile that would melt the knees of the strongest missionary. The young men were impossibly handsome, with thick brown hair, sinewy muscles and bright white smiles. This must've been the island where Calvin Klein and Victoria's Secret models are harvested. These people were freakishly attractive but not in the unnatural way portrayed on magazine covers at the grocery checkout stand.

Ian wallowing in Amazon quicksand. Note the distant bank. Photo by Jeff Wueste

The neat village was composed of generously spaced houses, with fenced corn and melon fields to keep out the free-range cattle, which seemed inordinately healthy as well, compared to the bony bovines of the high Andes. No one wore shoes, as the sandy soil was plenty comfortable and free of thorns. Children played and ran around safely wherever they wanted. Shade was plentiful, as was sunshine and a constant breeze kept the bugs at bay. I'd rarely, if ever, been to a more perfect setting during my travels.

Mozans noted the sudden change in the river at the Obidos Narrows, which has been called the Bosphorus of the Amazon. While the word "narrows" implies a thinning of the river, with the Amazon everything must be taken in context. Far from the hazardously tall cliffs and rapids of the Andean canyons, this narrowing of the Amazon still leaves more than a mile or two of water between the banks. When you're talking about a river that is normally five to ten miles across, a constriction of 80 percent is substantial.

Predictably, about a dozen ridiculously good-looking men and women hung around, watching us set up camp and prepare dinner. They laughed at our dehydrated meals and wisely declined a sample. For young, single adventurers making it through this area, an extended respite in this Eden would no doubt be beneficial to body and spirit. With a diet of fresh fruits, vegetables and fish and days filled with activity and rest in this perfect climate, this community had no choice but to be the picture of perfect health.

That evening Barbara relayed a message from Mauro, who could arrange a guide and boat for $250 a day, plus food and fuel. I told her that I wanted a firm price—food and gas included—to escort us five days from Santarém to Belém. I'd seen how this gas thing works, and it could end up costing us thousands of additional dollars as Erich found out on the upper Ucayali. I wanted a firm price, with half up front and half at the finish. If we couldn't make that deal, or afford it, then we were determined to go without an escort.

Chapter Twenty-Nine
The Riviera of the Amazon

Day 100, Mile 3,518

"Living on the road, my friend, was gonna keep you free and clean. Now you wear your skin like iron; your breath's as hard as kerosene."

Townes Van Zandt

Santarém was another significant milestone on our journey, marking the start of the final push to the ocean. Though the Atlantic was still 600 river miles ahead, the remaining distance was a mere fraction of what we already covered. In popular parlance you'd say that the miles had hardened us, but I found the opposite was true. We'd made it this far because we'd learned to be malleable. Our success was primarily due to our ability to adapt to one another, our environment and the conditions. We certainly weren't going to alter the mighty Amazon in the least, so the only path to success was to recognize, accept and work with the water, wind, supertankers and whatever other challenges confronted us. There was no battle to win. We weren't 'men against the Amazon' as so many unimaginative headlines tout. We were simply going with the flow and blending in as best we could.

Difficulties we faced were inherent and obvious, leaving complaints to the wind. Though we never set a hard and fast rule against it, none of us complained. If we let slip a discouraging word it was only a precursor to the solution that followed in the same breath. "These waves are huge!" wasn't necessarily a complaint, but an expression of astonishment at something new we were experiencing. Sometimes the exclamation was followed with a suggestion to move to the right or left to avoid the waves, which was my cue, as expedition leader, to issue a final decision about the course of action we should take, because I was the leader and much bigger, stronger, more virile, handsome and more intelligent than the other two. After issuing my stern and flawless directive, Ian and Jeff would ignore me and do what they wanted, and I'd follow them. It was a marvelous system. If I was lucky, they wouldn't laugh hysterically.

The waves upriver of Santarém, compared to the conditions thus far, were absolutely monstrous. There we were, minding our own business, skipping over the ocean like a song, when all of a sudden Toto wasn't in Kansas anymore. A switch had been flipped when we rounded the Obidos bend and for two days we battled strong headwinds pushing steep waves with faces measuring three to eight feet from trough to peak, with the occasional 10-footer to climb and slam down. The waves were bigger near the center of the six-mile-wide river, so we sought the relative calm nearer the bank. This was the narrowest section of the Amazon since well above Manaus and the result was like forcing water through a fire hose. The combination of the howling upstream wind, powerful river current and tidal surge rushing in from the Atlantic was like a Mixmaster blender stuck on frappé.

Ian's back and a standard Amazon hammock boat. Photo by Jeff Wueste

We left the Island of Swimsuit Models well before sunup to avoid the crowd with their siren song and, more importantly, to get a jump on the wind and waves.

An expedient process of elimination kept waking me up throughout the night, but the silver lining was that I got to see some of the most beautiful night skies on the planet after scampering down the steep muddy bank for an out of the way latrine. The wind howled up river throughout the night. Though it drove the mosquitoes away and made the sleeping pleasant, I had a sinking feeling the headwind would stay with us all the way to the Atlantic.

Sticking to the right bank around the big bend leading to Santarém, we paddled 14 hours without our normal break to push the 55 miles into town. Though it added a few miles to our route, taking the outside curve kept us out of the largest waves and, theoretically, put us in the fastest current. We needed all the help we could get against the wind. After dark, we kept a keen eye out for small boats running without lights and the mid-sized barges that stuck to the bank to avoid the ocean-going vessels that now owned the middle of the river. One after another, a procession of large container ships passed us en route to and from the Atlantic. Just to break up the tedium, Jeff photographed each one, even in the dark.

After the sun set, the wind eased significantly and the current was swift as we raced towards the lights of the city, still miles away. We were all much wearier than at the end of our normal days. With no break stop on land, eating and drinking on the run and pushing for hours against the wind, our bodies were worn and nerves frayed.

Ian and I carried on a running debate about whether the large bodies of water where tributaries joined the Amazon were lakes or merely the mouths of slow-moving rivers. Ian was for lakes; I argued rivers. Looking at the GPS, I zoomed in on the Rio Tapajos at Santarém, and showed the name to Ian. We all knew that *rio* meant "river" in Portuguese, but Ian was still not swayed. In the midst of this entertaining albeit pointless debate, Jeff and Ian were carrying on a hushed, but heated discussion that I couldn't quite make out. Their chatter ended abruptly when Ian took off in a pretty hard sprint.

"See if you can keep up with this!" he called out, in what were the first and last contrary words I've ever heard from him to this day. Ian was in the bow of the tandem kayak with Jeff sitting behind. Both were using their single blade canoe paddles at the time; otherwise, I doubt I could have kept up. I was using my double-blade wing paddle, which looks like a regular kayak paddle but has a scooped blade and is made from ultra-light carbon fiber. Their single-blades were carbon as well and resemble a standard canoe paddle, but with a bend at the blade for increased efficiency. We carried both types of paddle and switched between them to avoid fatigue and boredom. Now, Ian was a man on fire. His strokes were smooth and strong, though I wasn't quite sure what motivated him to sprint near the end of this trying day. The town was still five miles off and we had yet to even hit the wide mouth (or lake?) of the Tapajos, which we had to cross to get to Santarém. On we sprinted into the darkness, our vision obstructed by the bright lights of the city. After far too much of this effort my still damaged shoulder was starting to feel the strain I was trying to avoid. I called out to Ian, "Okay, I give up. It's a lake!" and we immediately and mercifully slowed back to our normal pace.

When I asked Jeff later about the episode, he told me that he was giving Ian some unsolicited instruction on proper single-blade stroke technique, which Ian wasn't in much of a mood to accept.

Once back down to a manageable pace, we let the silence dissipate whatever bad feelings had arisen. Finally, at the upstream edge of the Tapajos confluence, we discussed our game plan to find the Navy barge. After crossing the two miles of smooth black water to the waterfront the approaching lights of the city indicated Santarém was much bigger and more industrialized than we expected. Up the Tapajos, about a mile distant, two cranes suitable for loading ocean freighters stood tall and imposing. To this day, I have a deep, almost claustrophobic, apprehension about paddling in swift water near large industrial objects, be they floating or stationary. The city, though a welcome site, was foreign and for all their brilliance the lights were less inviting than the dark beaches we came to recognize as safe havens.

As we drew near to the waterfront, we found a boardwalk bursting with nightlife and laughing people dining and drinking on verandas overlooking the harbor. Large luxury yachts were moored just inside the Tapajos to avoid the swift flow of the Amazon. Weaving our way between yachts and wooden riverboats and following Barbara's directions, we worked our way about a mile up the Tapajos, where we found the Navy barge.

I called out a proper howdy-do in my best Portuguese, which sounds remarkably similar to loud English with a very bad Spanish accent. I felt, and probably sounded, like a North American version of Mickey Rooney's Mr. Yunioshi in *Breakfast at Tiffany's*. I climbed onto the barge, where a young recruit greeted me politely, with his hand on the service pistol in the back of his belt. I asked for the commander and with great ceremony presented the crumpled yellow paper explaining who we were and what we wanted, together with Cmdr. Fernandes' business card. We were quickly introduced to Sgt. Vlademir and Capt. Andrade. The Sergeant was a huge bear of a fellow, with a massive grin and happy demeanor. Capt. Andrade was the picture of military professionalism, quietly issuing commands where needed.

I've always viewed multilingual people as intellectually superior and hold them in great respect, regardless of their level of formal education. To me, this ability has always been well beyond my pay grade, in spite of best efforts. My attempts at Spanish entertained my father-in-law for years and has met with ramped-up eye-rolling by my adolescent daughter. If I were to have one super power, it would be the ability to read, write, speak and understand all languages. Until I'm bitten by that particular linguistically engorged radioactive spider, I'll continue to fumble through it all, with the most honorable intentions.

With a few words, the Captain and Sergeant made plans to secure our kayaks, haul us into town and drive us to our hotel, which Barbara reserved well before our arrival. No expedition should be without a Barbara. Sgt. Vlademir really enjoyed opening up the massive twin outboard monsters that powered the semi-inflatable patrol boat, as he shot back down the Tapajos to the city center. Moving so fast after being relegated to the speed of nature for so long was disturbing and exciting. We were met at the dock by a pleasant ensign, who took us directly to the Hotel Amazonas Boulevard.

It had been a long day, with a really nice ending that cleared away all the fatigue and whatever negativity the rough parts instilled. We all did our best to video chat with our loved ones back in the States, though Ian still wasn't able to reach Addie, which was heartbreaking to us all.

South of Dixie

"Whenever I find myself growing grim about the mouth; whenever it is a damp, drizzly November in my soul; whenever I find myself involuntarily pausing before coffin warehouses, and bringing up the rear of every funeral I meet; and especially whenever my hypos get such an upper hand of me, that it requires a strong moral principle to prevent me from deliberately stepping into the street, and methodically knocking people's hats off—then, I account it high time to get to sea as soon as I can. This is my substitute for pistol and ball."

Herman Melville, *Moby Dick,* 1852

At the end of the U.S. Civil War, about 20,000 people sailed from Texas, Alabama, South Carolina, Louisiana and Georgia to re-establish their way of life in countries that were more amenable to slavery. Though many eventually returned to the United States, several families established successful settlements in Brazil. Santarém was one of the towns where the *Confederados* established a secure foothold. Brazil was the last country in the western hemisphere to outlaw slavery, in 1888. Even then, it continued for some time in practice if not in law. The most well-known settlement is the town of *Americana* in southern Brazil, where the battle flag of the Confederacy is waved proudly and festivals are held, replete with rebel uniforms, hoop skirts, fried chicken, old-timey string bands and antebellum-style balls. Like many confederate sympathizers in the U.S., most descendants of the *Confederados* willfully ignore the enslavement of people by their forebears, touting "Heritage, not Hate" to minimize or negate the central role of slavery in the culture they revere.

The descendants learned English from their parents and grandparents, speaking with a distinct southern drawl. During his visit to the region, President Jimmy Carter made note of the accent, remarking they sounded exactly like people in his home state of Georgia. In Santarém, gravestones bear the names of Riker, Wallace, Jennings and Vaughn. Many more *Confederados* settled in Paraná and Belém.

We spent the next day in Santarém, organizing gear and trying to arrange for a gunman and support boat. Working through Mauro, we got news that Lt. Reboucas ("Heh-boh-sesh"), our contact for finding an armed escort, wanted $12,000 for the boat and four gunmen. Reboucas had been part of the security detail for the most recent Cousteau Expedition on the Amazon.

I countered with one gunman for $6,000, which happened to be the grand total of everything left in the expedition account. I reiterated that there would be no more money, even if our escort encountered unforeseen circumstances or ran over budget. There simply was no more money. The price was firm and it had to include all fuel, food, return trip and essentials for the boat.

After two nights of clean beds, air conditioning, showers, food and modernity, we were ready to depart. The comfort of Santarém only made our roughened habits, hands and resolve stand out in an uncomfortable fashion. I arranged for my mother to wire the remaining funds to us, which meant none of my family could afford to meet us at the finish. With plans to collect the money after lunch, we swam about 100 yards out to the Navy barge to collect the kayaks and paddle them into the beach to transfer our gear onto the hired boat, the *Confiança II*. It was a well-built wooden riverboat, painted white, with room to seat about 20 people. A strong wooden roof covered in sheet metal ran the length of the entire boat. The only enclosed spaces were the tiny pilot room up front, a shower room and the toilet out back. The *Confiança II* was powered by a straight-six gasoline engine beneath a box-like enclosure that doubled as a table. Neatly painted wooden benches lined the railings on both sides and orange life jackets were stowed in the ceiling. Hooks on either side of the ceiling were used to attach hammocks. The boat was moored to the shore with an anchor line deep in the sand, though the bow was beached in water only inches deep.

I pulled my loaded kayak onto the beach next to the *Confiança II*, where a gust of wind kicked it over and snapped the carbon fiber fin in half. I forgot to retract it when I got to shore as I had done automatically hundreds of times over the past two months. The wear and tear finally got to me and I, too, snapped. In front of Ian, Jeff and the world, I went completely nuts, yelling and cussing about the rudder and everything else that conspired to prevent us from getting on the water, kept me from my family at Thanksgiving and swallowed up the entire remaining budget. I railed about pretty much everything else under the sun, from diarrhea to crappy internet signals to the lack of decent coffee. As my tirade went into extra innings, Jeff and Ian slowly and wisely backed away, and found a reason to go to town for more provisions. The broken rudder blade was a straw heaped upon my back that was already loaded with the disappointment felt by National Geographic's abandonment of our expedition, personnel problems with John and Pete, Rocky's relentless public online attacks, constant weariness, the language barrier, lack of funds, missing my family, concerns about our safety, my damaged shoulder and the stifling weight of responsibility.

I was itching for a good old-fashioned bar fight. While I fumed, and patched my rudder on the beach, two older teenage boys came and stood near the tandem boat. They placed themselves much closer than is comfortable, the same way hoods stand a bit too close to intimidate in a plausibly deniable manner, on a deserted street. At first, I ignored them, doing my best to stew quietly in my rage. There was no sense involving these boys, who were too young to figure out they weren't as tough as they believed, in my turmoil. As I squatted down to work on the rudder, they moved in closer, not talking. One stood and watched while the other slowly walked closely around our gear and boats.

I wished, so badly, they would try something stupid, just so I could let off some steam. Please, please make a move, I silently hoped. Just a small bit of violence would take the edge off and I could move on. Surely, no one would blame me for beating up two teenagers. I mean, there were two of them. I figured that I was giving them more than a sporting chance. Perhaps they were bored or maybe they finally sensed the old guy was off his rocker, but they thankfully moved on without a word. I think they recognized batshit crazy when they saw it. They didn't deserve my ire and it was shameful for me to consider dispensing it upon them.

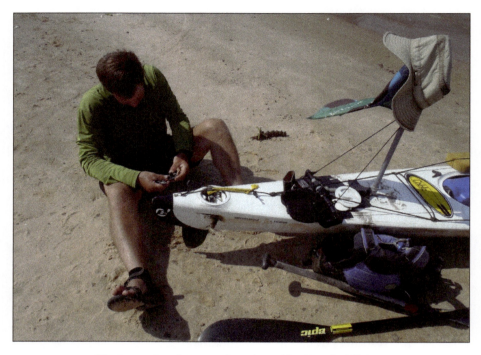

West in a mid-psychotic state fixing his rudder. Photo by Jeff Wueste

By the time my partners returned, I was back into the civil world and apologized for my outburst. Both guys were so compassionate and understanding as I explained my frustrations. I couldn't ask for better compadres, to put up with my shit like that. Our team was lucky that we traded off times to vent our anger. This simply had been my scheduled time for a blow up. Using duct tape, I splinted the rudder blade with pieces from a plastic plate and knife, though it meant the blade could no longer be retracted when running over debris or landing on a beach.

Lt. Reboucas met us on the beach, next to the *Confiança II*, to finalize plans. Oliviero Pluviano, a renowned journalist from Italy on location to film a documentary about the local Indios population, graciously helped translate. It took some work to get the point across to Reboucas that we only had $6,000 to spend, and simply could not pay more for any reason. I thought all of this had already been arranged and communicated through Mauro. Half the payment would be made up front and the rest when we arrived safely in Belém. Reboucas wanted us to pay for additional fuel for the return trip and additional expenses, should they arise. But I wasn't going to budge, nor could I.

Even after agreeing to my terms, Reboucas repeatedly came back to "and other expenses," for which I had to keep reminding him we weren't going to pay any "other expenses." Deep in thought, he agreed to work on the plan and get back with us the next day, meaning we were to stay another night in Santarém. Pluviano had clearly taken on more than expected and politely extricated himself. With thanks all around, we shook hands and went our separate ways, with plans for Reboucas to pick me up in the morning to get the money.

The beautiful city of Santarém is the "Riviera of the Amazon," located in what is known as the Savannah Trough, an area of particularly low annual rainfall, which extends north across the Amazon to the famous massif, Mount Roraima. The brick boardwalk is very neat and clean with lots of tourist activity overlooking ocean liners carrying passengers up and down the Amazon. While it boasts all the modern amenities, the cost of shipping goods to this remote location is astronomical. I spent $40 on a roll of duct tape and $100 on 16 AA batteries. These weren't even the gringo prices since they were on the price tags when I walked in the store.

Given the extra night, using the tiny tools on my Swiss Army knife, I disassembled the sat phone, dried it out and reassembled it to working order. Nothing stayed dry on the Amazon, though we were diligent with our electronics.

Capt. Andrade of the Brazilian Navy graciously allowed us to store our boats on his barge for another night. Each day we tried to leave and each day we returned to him, hat in hand, requesting his assistance. His paternal demeanor and understanding warmed our hearts. During one of our conversations, in what little we spoke of each other's languages, he proposed we invent a hat equipped with solar panels to charge batteries, because our solar panels couldn't get wet, so we weren't able to use them while paddling. The hat would be high enough above water to stay dry, he said. I'm sure the fashionistas would dig the look, too.

Oliviero Pluviano wallowing in far deeper than he planned negotiating with West and Lt. Reboucas. Photo by Jeff Wueste

Lawyers, Guns and Money

"I'm hiding in Honduras, I'm a desperate man
Send lawyers, guns, and money
The shit has hit the fan"

Warren Zevon

With our negotiations for an armed escort still uncertain, I spent the rest of the day trying to buy shotguns. I asked around in my broken Portuguese, but there was no word for "shotgun" in our little dictionary, so I used the word for rifle, *espingarda*. I inquired as to the permit. No one really knew. Then I asked several people where I could find a gun shop. No luck. Finally, the desk clerk at our hotel, who spoke English, just smiled and said we couldn't buy guns in Brazil. I disagreed, pointing out that we'd seen at least a dozen pirates on the river with guns. I told him that every five-year-old back home in Texas got a loaded gun in every box of breakfast cereal and legislators did their best to assure every serial rapist, child molester, fascist and wife-beater had unlimited access to firearms, as God himself intended. But not in Brazil, apparently.

At 8 a.m. on our latest "final day," Lt. Reboucas was supposed to tell us whether he would provide security within our strict $6,000 budget. Our remaining funds were actually substantially less, but my mother agreed to pay the difference using money she'd previously slated to fly to Brazil with the rest of my immediate family to meet us at the finish. I reminisced on the large sums of money wasted and extorted during the initial phases of the expedition.

Waiting for an answer from Reboucas, we discussed Itzhak Perlman, which led to Stradivarius violins, repairing them and such, then women, girlfriends, wives and ex-wives. Discussions about our children were cut short because it depressed us so much to be without them. Finally, I was able to reach Lt. Reboucas on the phone, and he assured me that $6,000 would suffice. Now, we just needed to get the money wired to us from Texas.

Lt. Reboucas drove us to the bank to get our money transfer from MoneyGram. Jeff and Ian waited outside getting an eyeful of the pretty gals, while the *tenente* and I went inside to deal with the money. When we got to the head of the line, Reboucas and the young woman behind the counter exchanged a lot of words, and then we were led over to another girl behind a desk. More words were exchanged.

Reboucas and the clerk both insisted I needed to show them my Siefpee. "What's a Siefpee?"

After some ineffectual back-and-forth, the lieutenant finally showed me his federal ID card and official number, next to the letters "C.P.F." The *Cadastro de Pessoas Físicas* is required for almost all official transactions in Brazil. Reboucas and the MoneyGram gal were incredulous that I didn't know about the C.P.F. as I bit my tongue on the whole issue of why they reversed the "P" and "F" when verbalizing the abbreviation. I was unclear how or why they would think that someone from another country would have an official Brazilian ID card. My official and highly detailed passport, recognized in every single country on the planet, simply wouldn't suffice. It wasn't a C.P.F. and I was obviously too stupid to understand any of this.

I explained that I wasn't a Brazilian citizen and therefore wasn't able to get a C.P.F., but I had a U.S. social security number. Again, I showed my Texas driver's license and U.S. passport. The lady typed my Social Security number into her computer, staring intently as if the Brazilian system would accept it. Shaking her head, she returned my Social Security card. No form of identification seemed good enough, and besides, we needed to go to the other bank down the street. Jeff and Ian followed in earnest as I filled them in on the debacle. I'm not sure, but perhaps they get a whole lot of US citizens with names similar to mine trying to retrieve wire transfers for a specific amount on this specific date in time. To all involved, at least on the Brazilian side, a Siefpee was strictly required and it was my fault for not having one. The fact that I was not legally allowed to procure a Brazilian ID wasn't their problem.

We got into the next bank and went right to a desk occupied by a pregnant young woman sporting a tattoo on her right shoulder. She was really nice and spoke some English. After some computer clicking she determined that the wire transfer reference number I had wasn't worth dirt. She did a search under my name. Finally, after more fast-talking between her and Reboucas, she told me that I can't withdraw half the amount in Santarém and the other half in Belém.

"What?"

Before that could really sink in, many more words flew back and forth between the *tenente* and the tattooed lady. Finally, I got a word in edgewise and explained that I didn't want a portion of the money now and a portion of the money later. I wanted all the money right now and that I never ever said anything about wanting part of the money now and part later. That part of the deal had been clarified the day before, when we agreed to give all the money to Reboucas up front. He would pay the *Confiança II* captain: half now and half when he returned.

Okay, she was clear on that, finally, but then said we'd have to go back to the first bank to get the money, since she only handled Western Union and this was a MoneyGram transaction. The Western Union office in Texas had given us this specific address in Santarém where they had sent the wire, which was a MoneyGram office.

Prying Ian and Jeff from their ogling, we went back into the first bank, where words were flown, kindly, between the *tenente* and the woman at the desk. After a few more words, we went back to the room where a long line of patrons still waited and settled into chairs against the wall. The first woman came up to the teller window, so we rushed her—against the grumbling of the folks in line—just so she and Reboucas could exchange another volley of words. Finally, Reboucas decided we needed to go to a MoneyGram store. Good idea.

We gathered the boys and walked several blocks to the MoneyGram place, where the young guys in matching MoneyGram polo shirts were glad to see us and confirmed that the reference number was indeed a MoneyGram number. However, the internet was down, so they couldn't give us the money. I asked if any other method could be used, such as the telephone or a dial-up computer modem. Nope. Sorry. Please, check back later. Reboucas got on the horn to a couple of folks, who confirmed that the internet was out all over town.

He suggested we hang out for a while near the MoneyGram shop in hopes that the world would tilt back upon its internet access. We spent an hour or so sightseeing and charging Ian's phone at a phone kiosk in the company of a young lady wearing lingerie in lieu of a shirt. Then we doubled back to the MoneyGram store, where the polo shirts gave us the thumbs down. Reboucas drove us back to the hotel and we checked in for an unexpected third night.

The evening seemed to take a turn for the better when we spotted a glossy brochure at the hotel touting El Mexicano restaurant. Home cooking! We hadn't had Mexican food since departing our home pastures. We made a beeline for the joint, which was playing Norteño music and decorated with sombreros and the like. The menu featured photos of familiar-looking fare, with descriptions in Portuguese and English. Tacos, the menu explained, were *"Mexican sandwiches, are corn tortillas on the plate-shaped crescent."* Under the heading for Snacks was a File Chimi Churry, described as *"Filet mignon meat baits."*

I called Lizet and left her a voicemail wishing her and Isabella a happy Thanksgiving. Isabella had a bad day and wasn't in the mood to talk. Teen angst ruled the day. Unfortunately, we had a bad sat phone signal and I lost her. The homesick knife dug in. My girls were having a great time with a huge meal at Sheila's house with family and friends, with turkey, dressing and all the fixings. Though we were missing the company and Thanksgiving feast, at least we'd have a taste of home with some straight-to-the-heart Mexican food. I was prepared to sit there and order meal after meal after meal, just to get my fill and go to sleep in engorged discomfort as Thanksgiving tradition dictates.

Jeff ordered the *Mexican sandwiches, are corned tortillas on the plate-shaped crescents* while Ian had the *filet mignon meat baits*. With great anticipation, I settled for what appeared to be chicken enchiladas with refried beans and rice in red sauce. A live band fired up some really great Brazilian country style music and thankfully avoided rehashing North American pop tunes. With a quick flourish, our full plates were laid in front of us, though a bottle of hot sauce was missing in spite of great efforts to explain how it was essential for the meal. Giving up on the hot sauce or any other fiery condiment, we dug in. Under the anticipatory gaze of my buddies, I was first to cut into my enchilada and relish the first bite.

This was not to be. The Porsche of food on my plate had a Volkswagen engine. It certainly looked good but wasn't going anywhere. The bland lump of mess looked as it should, but tasted like soft cardboard with flavorless cheese, drizzled with a line of ketchup. Jeff and Ian experienced the same anti-climax. Calling the waiter over, Jeff made note of the crap-tasting servings and asked what ingredients were used. The waiter called over the manager for Jeff to interrogate. The manager, in some pretty good English, explained that they had simply found the pictures on the internet and copied them as best they could with whatever ingredients they had on hand. For example, the red sauce on top of my enchiladas was in fact, Heinz 57 ketchup. He was quite proud of this method and related that this was a very popular restaurant amongst the citizens and tourists in Santarém. Jeff suggested he visit Mexico or Texas at some point in his life. We choked down the remainder of our disappointing meal, washed down with sufficient amounts of alcohol and soda, and departed much wiser.

Catch and Kill

On a phone call over breakfast the next morning, Barbara and I discussed the relatively pointed email she'd sent to the brass at National Geographic, asking them what, if any, plans they had for publicizing the expedition. She reminded them that we had lost many opportunities by keeping our end of the bargain not to speak a word about the expedition and the new source to anyone, though several media outlets had approached us. She further pointed out that we had missed out on fundraising opportunities because we'd stayed quiet about the new source. Mainly, she noted that we were let down by National Geographic's lack of any response to her request for additional funding. Not a "we're thinking about it" or a "no thanks" or even "go to hell." It was a tough letter, but not out of line. She received no response.

I agreed completely with Barbara's frustration, but told her we should not make any more efforts to correspond with National Geographic. We'd just wait for them to contact us. By now it was obvious there would be no cover story, and probably no story in the magazine at all. Yet National Geographic still wouldn't give us permission to speak with other media.

My initial efforts to gain sponsorship from National Geographic had gone nowhere, but that all changed when they learned about the new source of the Amazon. They moved quickly to gain control of the potential blockbuster story, approving funds for both my expedition and Rocky's, while securing contracts locking both of us out of any other media. National Geographic told us they planned to cover our respective expeditions in the magazine, and NGTV showed interest in the story as well.

The change came after Rocky unleashed his all-caps email excoriating National Geographic for its plan to cover both our expeditions in a single article, and then breached his contract to give *Outside* magazine the big scoop National Geographic spent upwards of $75,000 to secure. Following that, National Geographic seemed to pull all of their support for our source-to-sea expedition. It's understandable that they would want to cut their financial support, but to forbid us from talking to other media about the new source was completely beyond my comprehension.

While we were eating up miles on the Amazon, Piotr seemed to undergo a change of heart as well. Though initially supportive of Rocky's new source research and a great help to our expedition, he began shopping his theory that the Mantaro was not the "most distant" source of the Amazon, but rather an "intermittent" source. Piotr's new definition of a river's source gained traction at National Geographic headquarters, even though his reasons for discounting the Mantaro—the presence of dams and seasonal variations in flow—existed on all the great rivers of the world, including the Missouri-Mississippi, Yangtze, Murray, Volga and Nile.

Rocky wasn't helping matters by descending the Amazon on passenger boats, instead of finishing his scholarly article on the new source. It would be more than a year before *Area,* the journal of Britain's Royal Geographical Society, would publish Rocky's peer-reviewed paper on the subject. That gave Piotr, a legendary Amazon explorer with deep ties to National Geographic, ample time to cast shade on Rocky's claim with no one on hand to present the other side of the debate. The defining accomplishment of his extraordinary life, the first descent of the Amazon, was at risk and he wasn't going to let it go easily.

Immediately after *Area* published Rocky's peer-reviewed paper in February 2014, two years after the expedition, National Geographic effectively quashed the news. In an online story titled, "Where Does the Amazon River Begin?" staff writer Jane Lee yawningly described Rocky's research, starting with a sub-head seemingly designed to repel readers: "Five different tributaries have been designated as the source of the Amazon River through the centuries. A new study argues for yet another."

The big reveal, for which National Geographic paid more than $75,000, was dismissively reduced to "yet another" boring claim. The few dogged readers who persisted through the article were treated to a summary of Piotr's new intermittent flow theory, quoting him and three of his friends. Lee made no effort to reach Rocky, his co-author Nicholas Tripcevich, or me, though we were all mentioned in her article. It was classic sandbagging, and it had the desired effect. A handful of stories about the new source came out in the hours between the time *Area* published Rocky's paper online and the release of National Geographic's take. Afterward, there were none.

An Escape from Paradise

Right on time the following morning, Lt. Reboucas rolled up, ready to fetch the money. In a rare moment of facial expression and emotion, the *tenente* crossed himself when I asked whether he thought we'd have any more trouble at the MoneyGram office. The transfer went through with just one small glitch: the polo shirt behind the desk questioned why my middle name wasn't part of the MoneyGram order, but was clearly on my passport. I explained to him that it's not a common thing in the U.S. to use someone's middle name. He explained rather earnestly that in Brazil it is a common thing. I got on the phone to my mother in Texas, to which, she said there were only two blanks on the form for the recipient's name, so she logically wrote my first and last name. I explained this to the guy behind the desk, who simply didn't care. I rolled my eyes (huffing quite loudly, I'm not proud to say) grabbed my pile of documents and started to push away from the desk. The MoneyGram dude and Reboucas quickly calmed my irritation and told me it was okay. They'd give me the money, but in the future, make sure my full name was on the transfer. Right, I said. Next time.

Flush with cash, Reboucas drove me straight to his house to present me with three ornate ball caps with the insignia of his rescue-salvage dive team, an elite unit of the *Marinha do Brasil*. With a lot of short, loud words and hand gestures, we carried on a conversation of sorts about diving the Amazon. He described diving with manatees, which he said are quite furry. They tend to congregate where waters meet, just like dolphins and sharks. Yep, sharks on the Amazon, as if I didn't have enough to worry about. He showed me his Facebook page, which was full of photos of the giant anaconda he captured with Jacques Cousteau's men in a BBC special broadcast in the early 1990s.

After gathering the boys and gear, we went to the Naval base to say our goodbyes and thank Capt. Andrade. The Captain, in full whites, was very knowledgeable about the Amazon. We talked a great deal about the exact location of the end of the Amazon River and the best routes through the mangrove swamps. The Captain had a high opinion of Reboucas and Sergeant Dias (Geeesh), who Reboucas hired as our escort.

After lunch, we swam back out to the Navy barge to retrieve our kayaks, then met Sgt. Dias on the *Confiança II*. He didn't look like much and appeared to only have a 12-gauge pump. I was hoping for a uniform, pistol and rifle, at least. Reboucas introduced him as a "Miami Vice" type of police officer. He was wearing sandals, with a T-shirt and baggies over his gut and soft muscles. He spoke no English. We paddled out of Santarém at about 2 p.m. as Capt. Beto finished preparing the little *Confiança II* for the journey ahead.

Before we left, Reboucas took me aside several times to write down a list of expenses and the money I'd need to pay Capt. Beto in Belém. I did my best to clarify for the millionth time that I was prepared to pay the $6,000 I just gave him and not a dime more.

We paddled past the harbor and into the black water of the Tapajos River as it blended with the Amazon proper. The *Confiança II* caught up about three hours later. The escort stayed near the north shore in calmer water as we headed down the middle to take advantage of the current, then cut the corner on a wide bend and found a nice beach on the south shore just as darkness once again came too quickly. We tied up to the boat as it anchored and prepared to relax. Next thing you know Capt. Beto was firing up the engine and heading back north across the river. I freaked a bit and finally, with the help of the dictionary and one of the young crew, asked where the hell we were going. He said it was safer out of the waves across the river. I told him we need to come back to where we stopped the kayaks in the morning to start them in order to maintain consistency with the expedition. Safely at anchor on the north shore, I worked out a daily plan with Capt. Beto. At about 5 p.m. he'd find a good place to moor, then we'd meet up with him at 6:30 or 7:00 pm when we saw his lights.

Sgt. Dias, Lt. Reboucas, Ian, West, Capt. Beto and Jeff in front of the Confianca II (right). Photo by a friendly bystander

The two crewmen were Captain Beto's nephews, Joedson and Marlson Silva. Nice boys in their early twenties with easy smiles and a willingness to help. They really went out of their way to make sure our kayaks were safe and tied well to the boat. I spread out Debbie Richardson's sleeping bag atop the roof of the *Confiança II* in order to avoid the snores and cramped quarters on deck and fell into a blissful sleep to the gentle rocking of the Amazon.

Chapter Thirty
Big Water

Day 103, Mile 3,544

"If you can't fly then run, if you can't run then walk, if you can't walk then crawl, but whatever you do you have to keep moving forward."
                                                                        Martin Luther King, Jr.

Sgt. Dias sat around most of the time looking bored. We weren't sure if he was pitching in or just doing his job making sure pirates don't attack us. It was a pretty easy gig and I'm sure we could have accomplished the same by giving a couple of shotguns to the crew. His gut and lack of muscle tone instilled no confidence in his abilities, though he was a pretty big guy. Joedson and Marlson held the opinion that he wasn't a real cop. Still, as far as I could tell, he was the only one with a shotgun; therefore, he was the dude.

During the day, the Amazon handed us our asses. At night, the three of us ate our dehydrated meals pretty quickly and headed to bed as the crew cooked chicken and veggies. Joedson poked his head over the roof, where I was typing in my net tent, to invite me to dinner, which was awfully nice, but I was already full and plenty sleepy.

I tried to convey to Capt. Beto that we wanted to meet up for lunch each day (*"almoço reunião* doze" I said, keeping it literal: "lunch meeting twelve"). With equal parts lack of understanding each other's language and his lack of enthusiasm for the translation book, all communication was on my shoulders.

With the Atlantic Ocean coming within days, instead of weeks, the intensity of our homesickness grew. It was best not to think about it and concentrate on other things. Music was too distracting, so I built boats, remodeled our house, restored classic cars, imagined romantic moments, cooked meals, designed new kayak components and wrote this book, all while paddling. Jeff also built a house, so we compared notes.

Sunscreen, Jeff and Ian on the Confianca II preparing for another day. Photo by West Hansen

Needing more space from the crew, Jeff and Ian set up their tent next to mine up on the roof. The strong high tide started rolling inland around 6:45 p.m., right after stopping for the day. The impressive surge of water, while not the famous *pororoca* wave, was strong and fast. We hauled our kayaks up onto a muddy beach, where the *Confiança II* was tied. Joedson kept an eye on our kayaks, tied up onto the beach, to assure the tides wouldn't grab them and we could get some sleep. We were really worn out, after a dismal 54 miles of battling strong headwinds and waves. My hatches were holding up, which was good, but I wore down my heels again and had to wear my tennis shoes in the kayak. We warmed up to the crew, with the exception of Miami Vice (Dias), who just sat like a bump on a log, not interacting. The other guys, including the quiet Capt. Beto, were all very nice.

Nights were filled with a beautiful bright moonlight on the Amazon, which conjured up thoughts of a new cocktail, Amazon Moon: one part Amazon river water, four parts Johnnie Walker Red, three parts diarrhea, one part protein powder, four parts tuna, eight parts dehydrated food, one part pumped water, six parts Jack Daniels, twelve parts oatmeal, stir and serve with skinny greasy undercooked chicken with hairs, plantains and rice. Yum.

West and an ocean freighter 1000 miles from the ocean. Photo by Jeff Wueste

As we drew closer to the ocean, the river really started to open up. We could still see the northern bank but had to squint pretty hard at times to do so. Tides and wind worried the Amazon each day into full-sized waves and incessant chop, requiring all our attention to stay upright and pushing forward. Using our pee bottles while underway suddenly became a precarious maneuver. The landscapes were really beautiful though we saw heavy smoke to the south, where ranchers torched the jungle to clear pasture for cattle. The river itself was quite wonderful and it was a special treat to paddle lighter kayaks, with the *Confiança II* hauling our gear.

Joedson grabbed some fresh acai berries for our dinner. Note: if your acai is sweet, then A LOT of sugar has been added. Photo by West Hansen

The morning brought heavy wind and waves building to nine-foot faces. The strongest headwind we encountered churned the river into a frothing cauldron that crashed and spilled over the eight-foot mud bank to our right, sending huge plumes of water crashing into the air. It took everything we had to maintain two or three mph into the wind and waves. I couldn't let go of my paddle even with one hand long enough to grab a drink, so eventually I moved over to the tandem, holding onto its bow for stability while sneaking a couple of quick gulps from my bottle. The whole process took less than a minute, and though we were 100 yards offshore, when we started, the breaking waves were already threatening to slam us into the eight-foot sheer bank. Facing upstream for stability, I planned to turn around in a long safe arc to prevent getting broadsided by the huge waves. Jeff and Ian were able to back paddle away from the impact zone using their single blades.

As I scribed my arc toward the relative safety of mid-river, a huge wave gathered a head of steam and stood tall when it dragged bottom. The growing mound of water came at me from behind like a watery catcher's mitt, picking up my kayak and surfing me toward the bank in a way that would have made Laird Hamilton proud. I torpedoed through the water, standing on the right pedal to gain whatever leeway I could away from the bank with my barely repaired rudder blade. Back paddling, I barely stopped in time, and then swerved the bow around and sprinted clear just before another wave came crashing in.

Skirting up and quickly avoiding being slammed into the bank in huge wind and waves. Photo by Jeff Wueste

The *Confiança II* had to find its own route through the circadian torrent, frequently going far to the other bank or around islands to avoid the worst of it, as we just put our heads down and suffered through the melee. The clear blue skies and light puffy clouds stood in vast contrast to the storm-like conditions. On the upside, the boat crew started cooking for us, which unbeknownst to me had been part of the deal all along. Ian had long since given up on our dehydrated food and politely horded the instant apple cinnamon oatmeal for every meal, which wasn't near enough calories. The appearance of real food, with second helpings, perked him up like nothing else.

The Lower Amazon Grind

Day 108, Mile 3,804

"The real glory is being knocked to your knees and then coming back."
                                                            Vince Lombardi

As it nears the sea, the Amazon fans out into a delta like a giant oak spreading skyward. The main branches split, with most of the water flowing northeast into the Atlantic. A sizable portion splits off to the southeast, where it joins the Rio Pará and flows into the ocean about 70 miles east of Belém. Between these two main branches lies Ilha Marajó, an island the size of Switzerland. Dozens of smaller islands press against the larger island, with the Amazon's countless side channels spreading through them like veins and capillaries. These are known as *furos*, a word meaning puncture or hole. The tidal heartbeat keeps the entire coronary system in constant movement, save for a few moments in transition when forces of nature, far too large for any human to control, stop of their own accord and the currents change direction.

We chose the southern route because it is longer. Our goal was to follow the path of a raindrop that takes the longest-possible path to the sea. We started from the Mantaro because it is the most-distant source of the Amazon, and we followed the longest route to the sea for the same reason. This is the same logic Piotr and Joe used. They traveled the longer southern route too, as have most other Amazon explorers. We chose the right-hand channel at every split in the river, until at the end of the day we were in the Furo do Tajapuru, a narrow branch of the Amazon leading south toward Belém.

Smoke from burning timber hung thick as fog, burning our eyes and throats. We first noticed the fires between Tabatinga and Manaus, but they grew larger and more frequent as we traveled farther downriver and cattle ranches became more commonplace. Fenced barges moved herds of skittish cattle up and down the river, between pastures hacked out of ancient forests. We tied off to an empty barge, moored just offshore from a timber operation as darkness brought forth a beautiful full moon, the last we'd see on the Amazon. It shimmered orange behind the dense pall of smoke.

Across the ever-wider Amazon from the Almeirim Mountains, Brazil. Photo by Jeff Wueste

Cattle barge. Photo by Jeff Wueste

Maw traversed the Furo do Tajapuru in 1828 as he made his way to Belém to seek redress for the insult he endured at the hand of the Brazilian commandant in Santarém. The offending official had seized the Englishman's saber and arrested him on trumped up charges. Though later released, Maw had refused the return of his saber because the offer was not accompanied by a formal apology for the dishonorable circumstances of his detention. Maw took his position as an officer of His Majesty's Royal Navy quite seriously and, as an invited guest of the Brazilian state, he expected his hosts to do the same.

Nearly two centuries later we glided easily through the furo, chatting in our weary way about anything and nothing. This narrow cut reminded us of the Trinity River in east Texas, with its tall pines and dark water. The palm trees were the only plant we could accurately identify, leaving the mass of greenery mostly a mystery when it came to the topic of taxonomy. The morning wind died down and we made about 57 miles for the day. Despite the easier going in the protected furo, I felt an inexpressible overall fatigue. I was just plain tired.

Body maintenance was a constant issue. West patching up Joedson's cut foot. Photo by Jeff Wueste

It was nice to leave the cooking and choice of mooring to Capt. Beto and his crew. On the *Confiança II* I had a shower and shave, which I'd been avoiding like the plague. For me, it was difficult to get wet again after being wet with paddle splash all day long. I lost the fingernails from the little fingers on each hand from pulling hard on the wing blade, which wasn't uncommon in the canoe racing community. My jungle rot was improving somewhat, thanks to the gauze and duct tape wrapped around my heels before donning dry socks and damp running shoes each morning. I would have rather paddled barefoot, but my feet hurt too much. My right wrist was showing some wear, which I treated with liberal doses of ibuprofen.

At night, after wolfing down as many calories as possible, I sat in my mosquito net tent on top of the boat as howler monkeys crooned from across the Furo do Tajapuru. It was awfully nice to be in the 100-yard-wide furo, after spending months on a river that was at times six to 10 miles across. The jungle was magnificent, with fresh smells of green growth filling every available space. The odor of life and death mixed together in a thick mélange, playing in drastic contrast to the antiseptic state modernity strives for, and often accomplishes, in our cultural quest to unnaturally improve our lives. The plants look just as they should, quite satisfied with the moisture, river and space they could afford. For once we could see both sides of the river without turning our heads, providing a welcome break from the agoraphobia we experienced on the big river. The *Confiança II* tied up to a dock belonging to a church and school, all on stilts, where we received a warm welcome. Dias got right to work as we paddled into the narrow furo, staring down every passing boat to show he meant business. With his shotgun and holstered pistol, he struck quite the imposing figure, erasing any previous doubt we had of his fortitude and position. He was definitely a cop.

The air became more humid as we dove deeper into the furos, and the tidal flows grew far more powerful than we expected. At around three in the afternoon the tide shifted against us and our speed dropped markedly. If we stopped paddling we drifted backward.

Tiny channels along the banks of the furos are called *igarapes*, which roughly translates to "canoe paths." Such paths are kept cut and worn by paddlers traveling to and from their homes and work sites. Many led to lakes with beautiful beaches and great fishing, sheltered by palm trees. They also provided refuge for pirates.

We avoided the town of Brevis and aimed due south for the much larger Furo Santa Maria, which would lead us to the Rio Pará and the sea. The web of streams and rivers was accurately depicted on my GPS, which proved indispensable for our navigation through the watery maze. As we worked our way south toward the Furo Santa Maria the banks all but disappeared, replaced with mangrove swamps that held no solid ground.

Sgt. Dias in dialing it up to eleven. Photo by Jeff Wueste

Through the Heart of the Amazon

Day 109 (morning), Mile 3,905

> "Watching a coast as it slips by the ship is like thinking about an enigma. There it is before you, smiling, frowning, inviting, grand, mean, insipid, or savage, and always mute with an air of whispering, 'Come and find out.'"
>
> Joseph Conrad

The next day started earlier than expected when, from my perch atop the *Confiança II*, I heard our kayaks slamming against something hard. The wakes from the barges in the narrow channel battered our kayaks like frogs in a blender. Adding to the fury, the tide had gone out and the boats had less water in which to float. This was a huge concern, since my broken rudder was barely holding up. The noise woke me at around 3:45 a.m. and by the time I slipped on my wet shorts to hop over to the dock, Capt. Beto was grabbing the kayak, which bucked and pitched in the incoming wake. The cockpit was full of water, but otherwise it looked okay. We hauled it upon to the wooden dock where I could dump out the cockpit and survey the damage. Sure enough, the rudder was folded in half.

I looked up the word for "rudder" in the book and explained to Capt. Beto that I needed a "*novo*," which anyone with multi-lingual abilities knows means "new." Hell, it was four a.m. and I was trying to read a tiny translation book in the dark.

"Please help me build a *new*! I need a *new*! Can't you see that this kayak can't go on in these treacherous waters without a *new*?"

Jeff, having snatched the translation booklet from my hand, added the magic word "*leme*" and they all nodded in understanding. I don't think it was Jeff's interruption that caused the sudden epiphany, nor his use of the correct word for rudder, but rather my increased arm flailing that finally urged our hosts' comprehension. Arm flapping always helps get the point across.

I then showed Capt. Beto my floppy duct-taped piece of crap rudder and explained that a thin piece of *madeira* was needed. Once a piece of siding was found near the school on the shore, Joedson immediately took over. He cut a line in the board halfway through with our tiny folding handsaw, then with a small machete, thinned out a section to fit in the slot previously held by the broken rudder blade. He worked the machete like a surgeon. (Note to self: avoid, at all costs, any type of machete fight with a Brazilian.) We worked together affixing the new bombproof blade to the rudder assembly, using a bolt Jeff got in Santarém for just the occasion. The replacement fit perfectly. The only downside was that the rudder blade couldn't kick up when it hit an obstacle, so I needed to be careful pulling into beaches.

Joedson carving an exceptional rudder blade with a machete. Photo by Jeff Wueste

Marlson's handiwork got West to the Atlantic in fine form. Photo by Jeff Wueste

The day was fairly good, with no wind or waves. The downside was the incoming tide working against us, no breeze to cool us down, and the humidity dialed up to 11. We baked and sweated in the heat, which significantly affected our performance. Still, we pulled off about 50 miles. The furo was lined with brightly painted open-air houses on stilts, in groups of 30 or more, spaced about five miles apart. People smiled and returned our waves. Now and then, we encountered good-sized barges, pushed by tug boats, coming or going in the furo, just as a reminder of how deep and heavily trafficked the mangrove swamps can be.

We passed several small boatyards with projects of all sizes underway, each felled, milled and built completely by hand. One such yard, with several large boat skeletons in various states of growth, sported a sign in English, "Liverpool of the Amazon." The boats are built during the dry season on stands in the mangrove swamps, and then simply slid into the water when the tide is high.

Liverpool of the Amazon. Photo by West Hansen

Boatwright shop with piles of wood chips. Photo by West Hansen

We came across a German couple in a large galvanized metal-hulled sailboat coming upriver. There was no wind, so they had no option but to use their outboard motor. We talked with them for a while about the route and their journey across the ocean from Europe, then they gave us three cans of beer. I wouldn't normally drink in the middle of a hard workout day, but that cold beer was pretty incredible.

Teenagers around the world have a need for speed, and the kids living along the furos of the Amazon delta are no exception. Hot rod pecky peckys zoomed around and toward us in the afternoon. They were small, brightly painted wooden boats, no more than six feet long and a couple of feet wide—just big enough for one skinny teenager. The gunwales were barely tall enough to keep out the waves and the lawn-mower outboards propped on the sterns seemed gigantic in proportion. Like all hot rods they were meant to attract attention and go fast, their hulls adorned with flames, glittery lines and all kinds of racing stripes. They looked like surfboards with tiny gunwales around the edges. The unmuffled outboards deafened anyone within 100 yards and guaranteed permanent hearing loss for the young racers. The little two-stroke engines weren't any different than those mounted on the sterns of larger boats, but on these sleek little craft the result was sheer speed, the universal language of youth.

In 2015, three years after we passed this way, kayakers Tarran Kent-Hume and Ollie Hunter-Smart watched from their campsite on a dock as an armed group of these hot-rodders surrounded and robbed a passenger boat, then sped away. The tiny *igarapes* offer a quick escape for the speedy little boats.

Paddling is just part of life on the Amazon. Photo by West Hansen

The Rio Pará

Day 109 (midnight), Mile 3,919

"Spend sixteen weeks in the jungle and you being to question your own sanity, especially when you are the one goading everyone else ahead."
                                                                                                Tahir Shah

We'd started at 4 a.m. after a fitful night of sleep. It was threatening rain, so Jeff and I spent a quarter-hour rigging our rain flies as Ian snoozed. It rained all of

50 drops. We departed our dockside camp spot before 6 a.m., making good time until about three in the afternoon when we broke out of the furo into the Rio Pará proper. The tide was going full bore upstream. Our speed dropped abruptly from six mph to less than two. After about an hour of this we flagged down the crew of the *Confiança II* with plans to pull over for about four hours and wait for the tide to change, then continue downriver through the night.

Marlson joking around while the Confianca II is anchored in wait for the tide to turn. Photo by West Hansen

Conveying this amount of detail in the face of such language disparity was pretty much impossible. As we slow-talked loudly, the tide spit us back upstream at about two and a half mph. I urged Capt. Beto and his team to get the translation *dicionário* on the table. Instead, they just yelled louder in Portuguese, explaining why we couldn't put an eight-lug wheel onto a four-lug axle or perhaps something to do with pterodactyl umbrellas. Either way, I gave up any attempts to communicate and started paddling downriver toward a large island Capt. Beto said we couldn't use for camping because it wasn't *seguro*.

Watching me trail farther and farther behind Jeff and Ian due to lack of calories and the incoming tide, Capt. Beto relented. Using a combination of sign language and semaphore signals, he conveyed his intent to stop at the island, and then chugged ahead. Jeff and Ian went with him as my energy ebbed down to nothing. I was pissed at Beto for not stopping sooner and angry at Jeff and Ian for leaving me behind. They kept looking back but didn't stop or slow down. This irritated me for several reasons, most specifically because it seemed like a replay of the time John and Pete paddled ahead and spent the night on their own. In less than an hour Jeff and Ian were about a mile ahead.

Eventually, the *Confiança II* caught them and communicated something using, I don't know, sock puppets. They certainly didn't touch the goddamned book on the table full of words we'd actually understand. We all came back together in a cut between two large islands. It was well after dark and the channel was lined with well-lit houses and bustling with small motorboats.

I climbed aboard without saying a word. Beat down and pissed off, I didn't want to blow up at everyone. Our dinner was pretty sparse compared to the helpings dished out to the rest of the crew. Something had shifted with the crew. Ian explained, without me asking, that he and Jeff didn't want to lose contact with the escort boat so they just kept going as I fell behind. What the hell did they think, that the *Confiança II* would just take off without us if they stopped to wait for me? It was a ridiculous excuse and I didn't respond. More lessons in keeping my mouth shut.

After dinner, Jeff asked if I was going to talk with Capt. Beto about the idea of paddling during the out-flowing tide and resting during the incoming tide. Jeff sensed my frustration and took it upon himself to write down our intent and a list of questions for Capt. Beto in Portuguese. This seemed to work. Beto explained that the tide rushes in and out in four- to six-hour increments and the next ebb would start in about an hour. I was starting to feel a bit less irritable with food on my stomach, so I downed a rare cup of evening coffee and an even more rare second cup to get pumped for the night run. My mood improved greatly and I was glad I kept my mouth shut.

Before departing I spoke with Barbara about our plans. She was still waiting to hear back from Mauro and Lt. Reboucas about whether Capt. Beto would accept a Western Union payment for his services in Belém. He needed money to get back to Santarém, but Lt. Reboucas wasn't planning to pay him until our trip was complete. Just to clarify: From a small boat on the Amazon River, I used a satellite phone to call my sister in Texas. She then called Mauro in Manaus, Brazil to translate and relay a message to Lt. Reboucas in Santarém, who in turn called Capt. Beto, who was standing six feet from me on that same small boat in the Amazon River.

Barbara also began contacting news organizations about the end of the expedition. We were all pretty let down that National Geographic wouldn't publicly acknowledge or publish anything about our trip or the new source, but we had an expedition to complete, regardless. Barbara shared news of our nearly complete descent of the Amazon, but stayed mum about the new source, as we all did. Still, word was starting to leak out thanks to Rocky's contact with *Outside* magazine and his blog posts.

Under a cloudy, starless sky we started our night run with a 13.4-mile crossing of the southern branch of the Amazon, the Rio Pará. The wind and waves were tremendous, but by now we were feeling pretty cocky in our sea kayaks. As soon as we cleared the end of the island, Beto started to cross the giant river and we just followed his lights. The waves did their best to beat the crap out of us, but we had girded our loins with coffee, so all was well.

Now and again, we caught a surfing wave, but for the most part the swell was coming right at us. We took a significant amount of comfort in the glow of lights coming from the *Confiança II* and didn't have to read the river or find the route; all we had to do was stay with that little blip of light in that immense ocean of wind and waves. We just fell into a paddling rhythm that, after thousands upon thousands of miles, felt as natural as breathing. I ignored the GPS attached to my foredeck and simply worked the waves and paddled.

Once in a while an errant wave shot out of the night and I threw a low brace to stay upright. Because I was using a racing paddle with a scooped wing blade, I had to brace with the back of the blade. If I used the front of the paddle as I would on whitewater, the scoop would catch the water and pull me over. We were all honed and alert, working the wind, waves and water. This was our element. Our muscle memory and instincts adapted to whatever the Amazon threw out of the darkness. It was the devil we knew, and it brought out the best in each of us. My Epic 18x kayak launched over the waves and slammed into the troughs with a mighty boom against the dark water.

When we reached the south bank of the Pará, Beto tucked into the lee of Ilha do Caramujo and dropped anchor. Weary but triumphant, we tied on and crawled aboard.

As soon as we settled in, a heavy rain began to fall, so we dropped the tarps down to cover the sides of the boat. Ian elected to sleep on the roof in the rain, while Jeff and I slung our hammocks. I slept wonderfully beneath the protective mosquito netting of my Grand Trunk hammock. With everyone in his hammock and snoring loudly, huge cockroaches swarmed in massive herds from their lairs beneath the deck planks, making for any crumbs they could find. We were just 165 miles from the Atlantic Ocean.

The hours on paddle flowed with the rhythm of the tides, ignoring our natural circadian rhythms. As we made our way, the *Confiança II* made a run to the little town of Curralinho, returning with groceries and cold cokes just before the tide turned against us. None of us drank soda back home, but cold soft drinks were quite the treat in our depleted state.

I chugged my coke and settled into my hammock, listening to the Allman Brothers' Mountain Jam (Live at Filmore East) in my earbuds. We all zonked out for about three hours, and Joedson made us a hearty supper to carry us through our extended night run.

Capt. Beto directed us into a tight furo that snaked through the wide Ilha de Trambioca, allowing us to avoid the large waves in the bay. The smooth water and neighborly atmosphere of the furo stood in stark contrast to the conditions we'd encountered in Pará Bay. Sturdy open-air houses on stilts lined the banks and wooden motorboats plied the water. Cheerful Brazilian music emanated from houses and bars lit with bright lights and beer signs. Most notable were the Christmas decorations. Multi-colored strands of lights lined railings, doorways, roofs and balconies. Fake evergreen Christmas trees lit up in the windows of most of the houses. This was Christmas in the jungle and it was truly beautiful.

Chapter Thirty-One
Tidewater

Day 109 (evening), Mile 4,000

"If you don't master your circumstances, then you will be mastered by them."

Amos Towles

The crew of the *Confiança II* left us at sunset on the picturesque resort island of Cotijuba in the Baia De Marajó, where the Rio Pará meets the bay leading to Belém.

The bittersweet parting was made even more awkward by our lack of shared language. Marlson and Joedson pleaded with their uncle to follow us the remaining miles to the Atlantic, but Capt. Beto was determined to get his money and turn around. I figured there would be a problem with the money since our negotiations had played out like a high-tech version of the kindergarten game "telephone."

We were 10 days out of Santarém, and as close to Belém as we planned to get. Capt. Beto was smelling the barn, and understandably so. We pantomimed an offer of our last $200 to escort us to the finish, but even if he understood he wasn't interested.

We pulled the kayaks onto the beach on Ilha Cotijuba and loaded their hatches with all the gear we'd stowed on the *Confiança II*. It was a bit of a shock to see how much stuff we had, since we'd grown quite accustomed to the light, unloaded kayaks. It was like an old homecoming stuffing the baggage into the familiar spaces to prepare for the final 100-mile run to the Atlantic. While Joedson and Marlson helped carry our gear, Dias stared down at a handful of tough-looking characters who seemed to have taken an interest in our gear. In his lazy bear-like manner, Dias sauntered over to the lurkers, exchanged some pleasantries and made his imposing presence known. The pack of four guys quietly wilted and shirked away. Dias had indeed earned his keep.

Marlson, Ian, Jeff, Capt. Beto, West and Joedson parting ways. Photo by Sgt. Dias

We took photos and exchanged hugs with Marlson and Joedson, and then with some pushing, shoving and engine-revving the *Confiança II* backed into the river and turned toward Belém, which lies on a tributary about 10 miles from the Rio Pará proper. Once again, we were on our own but this time we felt different. Though we were tired and worn down we could smell the Atlantic.

I anticipated the end of the expedition for years. Originally, we planned to share the triumph with loved ones, but that travel money was now in the pockets of Reboucas and our *Confiança II* crew. We needed the security, but it stung knowing my family wouldn't be there to meet us, after all we'd been through. Jeff bought tickets in advance for Sheila and his grandson, and Jason accompanied Erich to capture the end of our quest. Erich planned to hire a speedboat to follow along and photograph us, but when we reached the Pará he decided to charter a helicopter instead.

The welcoming committee came together in Belém in a tragicomedy of errors and missteps. Sheila never spent much time around kids, so the intercontinental trip with Jeff's 12-year-old grandson, Gabriel, strained at her mooring lines a bit. Gabriel was a natural traveler with a boy's sense of humor, so he was great at every turn. Sheila failed to find the humor in his hijinks and couldn't match his roll-with-the-punches approach to traveling. Having found the hotel in Belém unsatisfactory, Sheila moved herself, Jason and Gabriel to a new hotel in Mosqueiro, a suburb along our route to the sea. Erich stayed in Belém looking for a helicopter.

At this point in the expedition, I didn't want to see anyone from the gang of four, as any meeting would just delay our push to the Atlantic. They had nothing to offer us that would expedite our process, which meant they would hold us back and further drain our dwindling energy. I was tired and impatient, and my family was still in Texas.

Having packed the kayaks, we paddled east toward the sea as the sun slipped below the horizon behind us. Ahead lay the seven-mile crossing to Mosqueiro, which would take us across the main shipping channel for vessels coming to and from Belém's busy harbor. We prepared for crossing in the dark, with heavy swell hitting us broadside from the left, by securing our cockpit skirts, eating and drinking. Within minutes it was full dark and the huge metropolitan area of Belém revealed its true extent. As far as we could see to the south, circling up to the north and in front of us across the bay, was a constant flow of street lights, buildings, moving traffic and radio towers.

Founded in 1616, Belém (Portuguese for "Bethlehem") is formally named Santa Maria de Belém do Grão Pará, though it was known simply as Pará until the early part of the 20th century. The city is renowned for its indigenous mango trees and extensive flora. Herndon noted several species, including the Paracide tree, which grows like a creeper around an existing tree, strangling the trunk, eventually cutting off nutrients and killing it.

A handful of huge ships dotted the blackness and barges passed in front and behind us, often too close for comfort. The GPS showed a lighted dinghy a few miles ahead, which we spotted and used as a visual reference to hold our course while fighting the sizable waves rolling down on us. Not long after dark, as if we didn't have enough to deal with already, a heavy fog moved in. Our concern about getting run over by a barge or fishing boat in the darkness and fog was only slightly overshadowed by the sphincter-tightening effects of the massive tide rushing in from the mouth of the Amazon. Ideally, kayakers turn into waves to hit them head on, which is far more stable and gives the paddler a fighting chance should the wave break. However, to maintain our course toward Mosqueiro, we had no choice but to take the waves from the side. Turning into the waves would have us paddling out to the middle of the river, which was miles from shore and vulnerable to even more wind, waves and ship traffic. So, on we went, bracing as needed and pumping the paddles hard. At one point, with Jeff and Ian to my right, a particularly large swell swept me up sideways, nearly causing my boat to slam into theirs like an oversized bowling pin. A quick brace was the only thing that kept me upright. That waker-upper kept us all on our toes.

We were in a constant state of apprehension in the dark, despite the ever-widening horizon of city lights and traffic spread along the shore a few miles in front of us. The crossing, though relatively short, took the better part of two hours due to the challenging conditions. With the diminishing waves hitting the shore, we surfed into Mosqueiro, beaching at a large fishing pier near a plaza. My wooden rudder blade, which was fixed in the down position, necessitated me jumping out in knee-deep water before dragging along the shallow shore. Bone-weary but wide awake, we quickly decided to press on and find a more suitable camp spot away from town. Earlier that day, Sheila told Jeff about a commercial campground just downriver from the Hotel Do Farol, where she and the rest of the ground crew planned to stay. It wasn't marked on the GPS, so we were to follow the shoreline until we found the campground or another suitable spot.

Jeff watched the boats while Ian and I walked to the town plaza to resupply with bottled water, Guaraná soda, hot sandwiches and M&Ms. Despite the late hour, just before midnight, the activity in the plaza was just starting to heat up, with a well-dressed crowd gathering in the humid, but oddly comfortable weather.

By the time we got back to Jeff, the tide was rushing into shore. What we thought was an incoming tide as we crossed the bay, had actually been neutral, which caused us to wonder what the in-rushing tide would be like. Jeff wrestled the boats up the beach a bit, but both were easily floating and needed to be held in order to keep from tumping over in the growing waves. We made a quick call to Barbara to alert her of our status and plan, stowed the supplies and hit the water. The GPS showed several large rock outcroppings, which in these waves could easily damage our boats. In the darkness these hazards were invisible. Now and then, Ian would come over to take a look at the tiny GPS screen and offer an opinion, which was good, because he was usually right in correcting our trajectory.

After getting a safe distance from the breaking waves we handrailed the shoreline to our right. The lights moving along the highways and roads were disorienting; it had been months since we'd seen traffic. Paddling too close to shore left us vulnerable to breaking waves, so we had to stay just outside the breakwater, which made it difficult to scout the shoreline for potential campsites. Eventually, we came to a spit of rocky land jutting out from the shore, with an old lighthouse perched atop a small island a couple of hundred yards out. A dark area surrounded by large trees was evident just upstream of the lighthouse, but the breakers and rocks were treacherous. As we contemplated the risky landing, someone appeared out of nowhere waving a flashlight. We cautiously headed in until we were shocked to hear Sheila's voice directing us back out and around the lighthouse. After getting past the rocks, we easily paddled the placid water in the shelter of the rocky dyke.

There on the sandy beach, grossly out of context, were Sheila, Gabriel and Jason. The last we heard, they were trying to find a hotel in Mosqueiro, with no possible plans to meet us. Now, here they were all bright eyed and bushy tailed. We had inadvertently paddled straight to the Hotel Farol, where they monitored our crossing via the transponder. My concern about meeting them melted away with their enthusiastic greetings. Besides, we'd be able to dump unnecessary gear with them.

We were all eager to push the remaining 70 miles to the Atlantic, so I didn't want to linger. We planned to wait just long enough for the tide to turn in our favor, and then press on. That meant two hours of sleep for Ian and me, while Jeff went upstairs to the hotel to rinse off and bed down with Sheila, which was cause for concern. I figured it would take a team of horses to pull me away once ensconced in the warmth of femininity, and I doubted Jeff would get moving under similar circumstances, especially given his notoriously gradual starts even on the best mornings. I voiced my concerns in the most diplomatic terms I could muster. After a few minutes hacking out the day's events on the laptop, I squeezed in some earplugs to block out the loud Afro-Brazilian beat from a party down the beach and stretched out on the tent floor. My alarm chirped in what felt like one second later. It was 1:45 a.m., and before I could roll out of the tent, Jeff strolled up, ready to go. Jason helped us pack up and then carried all our extra gear to his room.

The Incredible Heaviness of Being

Day 110, Mile 4,054 (Mosquiro)

"Roald Amundsen planned for everything to go wrong, while Robert Scott planned for everything to go right. If you expect a flawless expedition, a perfect occupation, a well-behaved child or a fairytale marriage, you will fail them all."

West Hansen

The tide came in and was due to turn soon, so we quickly shoved off into the darkness. It was a lonely feeling, buffered only by the company of my friends who shared this journey across a continent. Originally, the ground crew was to hire a boat to follow us, but they couldn't find a captain willing to go out in the rough conditions.

Rock outcroppings, derelict structures and the wrecks of sunken ships littered the bay. I couldn't imagine attempting this stretch at night without the help of our GPS. We kept our headlamps switched off, preferring to trust our night vision. Even with our eyes fully adjusted to the darkness and our senses at high alert it felt as if we were walking blindfolded through a minefield. Several times I called out for the team to zig left or zag right to avoid a hazard marked on the GPS screen though invisible to us on the surface. Small fishing boats cruised back and forth outside the breakers and beyond, setting out huge nets with plastic soda bottle floats.

Though we'd slept for two hours, we were now in the third day of an extended push for the finish, and the accumulated fatigue and lack of sleep left our bodies weary and minds clouded, even after an infusion of coffee and oatmeal. We struggled to stay alert as this stretch of water was no place to nod off, especially in the dark. For the next several hours of the night we followed the undulating shoreline, spending a great amount of effort dodging the submerged and partially submerged rocks depicted on the GPS. I scraped bottom just once, when I mistook a small island ahead of us for the rocks under us. I had to constantly zoom in and back out of the GPS screen to check our position in relation to the shore and that of the smaller rock hazards. I relied completely on the little glowing screen, at one point insisting that there was no peninsula ahead of us despite the house and car lights marking its outline. Ian tried in vain to convince me to trust my eyes, but

I didn't believe it until I zoomed out the screen far enough to see the peninsula, two miles ahead.

As morning approached and fog rolled in, Ian insisted we follow the shoreline, which according to the GPS took a hard-right turn at the east end of Ilha Mosqueiro. Ian's route, if we stuck with it long enough, would have taken us into the Baia do Sol, which eventually led back to Mosqueiro. After a lot of checking and re-checking the GPS Ian finally agreed to leave the comfort of the shoreline and follow a bearing into the disorienting thick hazy void.

Across the darkness to the north, too far for us to make out lay the easternmost tip of Ilha Marajo, jutting out into the Amazon delta. In 2019, the body of a 26-foot-long humpback whale was discovered 50 feet into the jungle from the shore. The island sits approximately 60 miles from the Atlantic Ocean, and Araruna Beach (near the town of Soure), where the whale was discovered was another 20 miles inland. Scientists surmised the calf was sick from consuming plastics, then drifted in on the huge Atlantic tide, once separated from its mother who could not navigate the shallower waters of the delta, then was thrown upon the island by the huge waves generated by a storm.

We slogged through the dull pre-dawn hours, occasionally dozing or nodding off as we moved through the smooth dark water. I was envious of Ian's lone position in the bow of a tandem boat, which I knew through my own experience in the same position during races. Jeff had to stay alert to operate the rudder pedals while Ian, in the bow, could shut his eyes for long periods of time without harm. These snoozes had become part of our daily routine, with Ian shutting his eyes for a few minutes at a time, and occasionally for as much as two hours, while maintaining a healthy pace. I envied Ian's naps but couldn't begrudge him. I'd do the same if I could, and it wasn't as if anyone could accuse Ian of not pulling his weight. Now as we paddled through the dark and fog I'd oftentimes startle awake with my kayak pointing off into the wrong direction and the tandem boys following dutifully behind. There was no way to be subtle about the heading correction so usually I just confessed to falling asleep at the wheel.

Sunrise ever so slowly inched through the fog, revealing the shoreline ahead and to our right. Waves were comfortably low, though shallow areas were made evident all around us by small breakers. A few rock islands where easy to avoid in the daylight, though the extent of the submerged rocks surrounding the small islands often came as a surprise. Small fishing boats were everywhere, spreading their float nets that extended for half a mile or more. The tandem kayak could pull up its rudder to safely paddle over the nets, but my fixed rudder would snag on them, requiring Jeff or Ian to free me. We did our best to avoid the nets but the floats were difficult to see until we were upon them. The most popular fishing area lay where Rio Guajará spilled into the Amazon River. In the morning light, we made out swaying palm trees and pristine beaches, and the occasional villa with well-tended landscapes stretching to the riverbank. This area, like many of the fresh water beaches of the Brazilian Amazon basin, rivaled anything offered up by Caribbean islands. To the north, all we could see was miles and miles of endless water. Somewhere, around 90 miles away, lay the northern shore of the Amazon River.

Soon after a blinding sunrise, into which we had to squint, a strong wind built up from the east-northeast, driving whitecaps that hit us just to the left of straight ahead. A direct headwind would have been easier since the crossing waves made it difficult to hold our course. Rudder-scraping shallows extended about a mile from the shore to our right, and several times we had to get out and drag our boats to the deeper channels between ankle-deep sandbars. When we ventured into deeper waters, large breaking waves battered us from the side, hindering our progress. With over 4,000 miles of trials we overcame to get to this point, the Amazon wasn't going to make this an easy finish. By now, we were happy to be doing three miles per hour. On we plowed, debating the best route, checking and rechecking the GPS. Ahead I could see what I was sure was a peninsula blocking our path to the Atlantic; in fact, it was the shoreline merely curving north at the mouth of the Rio Guajará. The immense size of the river and our proximity to the southern shoreline was difficult to process. There was no doubt we were on the right track as the GPS had been dead nuts accurate since departing Manaus. It was just a question of how much longer we had to fight these conditions in our worn-out state.

When we hit "the false bend" around noon, we all agreed a brief lunch break on shore was in order. Jeff spotted a picturesque spit of sand with the requisite palm trees a mile or so to our right. This idyllic spot lay ever so slightly in the wrong direction. Ian and I, not wanting to take one single stroke that wasn't leading us to our goal, vetoed the landing. We figured we'd find another stopping point soon enough.

Turning to the north and sticking closer to the shore, the wind and waves grew significantly rougher. Finally, we pulled onto a flooded piece of shallow mud near the shore, but only stayed a few minutes, as the waves threatened to beat the kayaks against the trees. The shade was heavenly and hard to leave. I was in constant worry about my fixed rudder scraping against the sandy river bottom. Not that it would break off, but would jam up and damage the hull or hang me up where I couldn't get out or move. My light solo kayak had an advantage over the heavier tandem, easily riding over the waves instead of plowing through them. Being rock-solid in their tandem, Ian and Jeff stuck to single-blade paddles.

The going was brutally slow and difficult. The waves constantly pushed us toward the flooded mangrove swamps, and we'd have to turn sharply and paddle hard to stay away from the shallows. The larger breakers admonished us the second we paddled out too far. Now and then a small trawler would buzz by to wave or stare at the aliens, which led us to question why our land crew didn't hire one of these boats to follow us to the sea. Several small fishing boats shared the waters through which we sought the sea.

After we crept another five miles in three hours, we found a tiny crop of land poking out from the mangroves. It was only big enough for the two kayaks, but it had shade. Every inch of exposed skin was wind burnt and sun swept. As soon as we pulled up and stepped out, our feet sank in the crusty mush. The entire semi-solid peninsula was covered in tiny barnacles. Even the tree trunks were encrusted up to about four feet. It was pure joy to drag up the kayaks, do some light maintenance, eat, drink some coffee and stretch for a bit. I lay back in my cockpit and shut my eyes. Despite the hot day, I found the hot dehydrated food pack and cup of coffee quite welcome. I suppose we were all pretty calorie deficient. The weariness I felt early on in the expedition seemed to have finally caught up to my partners. If not for our inner auto-pilots, we surely would have succumbed at this point.

By 2:30 in the afternoon we'd been at it for almost 12 hours without a break. We had to ration our drinking water, as the Amazon was too brackish this close to the sea to effectively use the filter, as we had throughout the trip. This rationing left us even more dehydrated than usual as we realized in our midnight stupor that we departed with less than enough drinking water. I took this opportunity to unwrap the sat phone and check in with Barbara. Erich finally secured a wad of cash substantial enough to get a helicopter to come out to catch us in living color, *a la* Magnum P.I. The problem was that he still didn't actually have the cash in hand and chopper pilots in Brazil operate on a pay-first basis. Erich wanted us to wait for a day while he arranged the flight. Of course, she didn't have to check with me before telling Erich there was no way in hell we were going to wait for him or anyone else. She knew we were ready to finish this sonofabitch, whether or not it was captured on film. Erich and Barbara bounced some tense words off a satellite, with the end result being that Barbara agreed to pass his request on to the team, which she did now as we wallowed in ankle deep mud in the mangrove swamps of the Amazon delta.

Our intrepid ad hoc photographer (foreground) and the motley crew on our barnacle encrusted mud haven. Photo by Jeff Wueste

At Jeff's suggestion, we took a vote and democracy reigned: love him as much as we did, Erich could go to hell. At this point, with life torn away, with our bodies grasping for whatever fuel available, with missing those we love, with the wind, waves and river working against us, dehydrated and weary beyond words, stopping was simply not an option.

Erich covered my ass many times before and since that moment and I loved him like a brother and felt his exasperation, but all of this took a backseat to getting to the goddamned Atlantic Ocean. For our team, this was more than a photo-op for National Geographic, which had already written us off and obviously didn't give a damn about the hell we endured. We were going to finish this thing, right fucking now.

Barnacles cover everything in a crunchy mess, showing the elevation of the tide. Photo by West Hansen

Within 30 minutes, our tiny mudpit of respite was flooded by the rapidly rising tide. The Atlantic spilled forth like an opened dam. We quickly packed our gear and were back on the water, facing even heavier wind and waves. Normally, we took a three to five-hour nap on the moored *Confiança II* during the incoming tide, but there was no place to stop in the flooded forest. Waves rushed through the mangrove swamps, crashing into tree trunks and tangles of driftwood and vines. Near what was the closest thing to being described as a "shore," dead branches, trees and stumps poked dangerously up through shallow water.

Whenever we could peer through a break in the thick woodland, the water went on as far as we could see. I'm sure there were high spots and eventually solid ground, but we would have a heck of a trip through the jungle to find it. Still, there were a few sparse signs of modernity in this otherwise prehistoric setting. We passed a few desolate radio towers and old concrete platforms, long since eroded down into the mud. This was an area constantly exposed to the Atlantic and nothing, natural or manmade, was permanent against the incessant onslaught.

Jeff and Ian steered their boat in a zigzag pattern, diving into the breaking waves head on for stability, while I was able to maintain a straighter line with my shorter kayak, surfing the waves for a while, then breaking out again as the waves petered out. We danced in and out of each other in a lonely manner, unable to talk or communicate in the high winds. Now and then, we paddled near one another until a large wave threatened to bash us together, whereupon Jeff would peel out to avoid the collision. He was irritable and responded by paddling far away, which left me feeling even more alone. Ian trudged on in silence as he was wont to do. Talking was superfluous.

Though we hadn't seen any dolphins since the Belém area, the bird life had changed. Most notably, we marveled at the egrets. They were bright red, white and blue. Not each bird, mind you, but individually solid reds, whites and blues. The red egrets were like moving Christmas lights. When flying in multicolored flocks, the colors would stay in the same groups while still being part of the larger flock, making wondrous living art. Birds of a feather, as the saying goes.

A few more hours of fighting the waves necessitated a much-needed break. The tall waves hitting them from the side had taken their toll as the boys wrestled with the larger tandem boat. I tried to tether my kayak close to shore in the hodgepodge of exposed debris, but the barnacles cut my hands and scraped against my hull like a cheese grater, so I headed back out. Eventually, when we could find no suitable shore on which to beach, I plowed straight into the jungle and found a patch of knee-deep water. I motioned for the guys to follow, and they rested a few minutes in their seats while I waded around and cut off a piece of my wooden rudder that had jammed up into the hull, preventing me from steering properly.

After some snacks and a few precious sips of water, we backed out as the Atlantic rapidly sucked the bulk of the Amazon out, like a planet-sized bathtub drain. I barely made it back out to the deeps as my rudder dragged bottom through the area I easily traversed just minutes before.

The next few hours we inched ever closer to the Atlantic, steering around giant wooden fish traps that resembled horse corrals rising out of the water. Fish camps and empty huts on stilts appeared along the newly exposed beaches to our right. With the huts came more fish traps, with their tall spike-like posts threatening to impale our delicate hulls with every rising wave. We wove through each fence carefully while balancing and surfing the low waves. The tide dropped precipitously as we reached Point Taipu just ahead of the setting sun at our backs, where two of the four previous Amazon source-to-sea expeditions ended their journeys.

We paddled on, determined to adhere to a more exacting standard. Just as we'd sought out the most-distant source of the Amazon from which to launch our expedition, our goal was to continue all the way to the ocean with no room for doubt or ambiguity. That meant applying a definition that is clear and applicable to all rivers. Simply put, a river ends where the ocean begins. But in a vast delta that includes myriad islands or marshlands—such as the Mississippi, Ganges or Amazon—where is this line of demarcation? David Kelly, Piotr and I had discussed the endpoint over the past two weeks and came up with a universally applicable definition. We took the two points on each side of a river's banks which extended farthest out to sea, then drew a line between them across the mouth, marking the end of the river. This is also how Piotr and Joe determined their finish point in 1986. David gave me the coordinates where Piotr and Joe had finished, and I punched them into our GPS unit. I labeled the point, "Stop, Forrest!" in homage to Mr. Gump.

At Pointe Taipu, "Stop, Forrest!" was still more than 12 miles away.

It was a bit after 5 p.m on December 4th and the zombie state was upon us. Still, we made forward progress and the GPS slowly ticked off the miles. Small fishing trawlers crowded through the deep channel separating Point Taipu from a huge sand island that stood about 12 feet out of the water, the highest scrap of terra firma we'd seen all day. Just then, like a scene out of *Apocalypse Now*, a helicopter zoomed in from behind us at treetop level. Erich found his ride.

With our very own Dennis Hopper hanging out of the circling chopper, we waved and took pictures of him as he waved and took pictures of us with a monster-sized lens that would have given Sigmund Freud pause. It was an exciting and uplifting moment after such a hard day. All the fishermen were waving and pointing at the shiny new helicopter as it circled. After Erich's chopper departed, we talked about all the different areas we would have preferred Erich photograph during our trip through Brazil. No matter; with the photo shoot duly crossed off the list, we paddled straight to the big sandbar across from Point Taipu and set up camp.

Approaching Pointe Taipu, with 12.5 miles remaining to the Atlantic. Photo by Erich Schlegel

*Near the mouth of the world's longest and largest river. Opposing bank is 90 miles away.
Photo by Erich Schlegel*

After 40 hours of paddling with only two hours of sleep and very few calories, we simply had to stop. A strong wind kept the mosquitoes at bay while we secured our tents into the soft sand and washed down a few peanut M&M's with sips of water, which was now too precious to use with our dehydrated meals. Across the small channel separating our island from Pointe Taipu, tall trees swayed in the wind against the lofty metal tower on top of which perched a light and radio beacon to warn incoming ships. The steady march of fishing boats through the channel signaled the end of their long workday, which began before sunrise. Within an hour or two their pocking diesel engines would cover the distance that took us 15 hours of hard paddling.

West accurately pointing to the direction of the Atlantic Ocean at Pointe Taipu. Photo by Jeff Wueste

We were elated to find the high ground, even though it took a lot of effort to haul the boats up the 12-foot sandy bank. Across the channel, trees and foliage covered the shoreline in every shade of green, while birds of all feather darted about, searching for a good resting place for the night. The mangroves kept any roads or development away for at least 20 miles inland, so the point remained much as it has been for the past few thousand years, save for the light tower. Our five-acre island was bereft of even the smallest greenery, though the ubiquitous gulls scrounged near the shallow rivulets that spider-webbed around our little encampment. Hardly a word passed between the three of us as we went through our well-oiled rituals, the months and miles having worn us down to only the most essential effort to exist. In the face of such exhaustion, there was a comfort in our routines and an even greater comfort in our camaraderie. I set my alarm for 5 a.m. to catch the outgoing tide, then with the wind buffeting the nylon walls, I fell back into my tent as the clock tipped 7 p.m., grateful for the full night's sleep to come and a mere 12 miles to complete tomorrow.

Chapter Thirty-Two
The Maw

Day 111, Mile 4,103.2

> *"I was alone and not that alone-with-a-book-on-Saturday-afternoon kind of alone."*
>
> Astronaut, Michael J. Massimino

Every single morning of the expedition Jeff needed close to an hour and at least one cup of coffee to get out of his tent. This routine wouldn't even begin until Ian or I sang him a soft wake-up tune and placed a steaming cup of coffee in hand as we gently cajoled him to sit up and crack open his eyes. So, there was more than ample reason for me to be concerned when none other than Jeff himself shook me awake at 8:40 p.m. I stumbled out of my tent door into ankle-deep water. Jeff and Ian had been awake the entire time, watching the water rise as I snoozed. The Amazon had risen 12 feet in less than two hours with the incoming tide. I couldn't imagine the volume of water it took to fill the Amazon's 200-mile-wide mouth to such a depth so swiftly, nor did I have to imagine it as endless dark seawater engulfed my world.

For miles around all I could see was a dull black void and water rushing in from the Atlantic. Save for the dull beacon atop the Pointe Taipu steel tower, not even a distant light from a lone village or a single star shown through the shapeless dark world inside which we now inhabited. During my nap, the wind died down from a howl to a steady pleasant breeze, but it still came from the Atlantic—a headwind. The salty green lush odors that filled our sinuses just hours before were wiped clean, leaving a sterile nothingness broken only by the occasional sharp wisp of our own stench whenever the breeze failed to carry it away. The loneliness of our situation pressed upon us more fully during each brief reprieve when the wind died and the interminable silence hugged us from everywhere, as if it waited just beyond our senses to remind us over and over, lest we somehow forget, that we were an eternity away from anyone who loved, hated or simply didn't care a thing about us. Then and there, we three were the only people in existence.

All that remained of our five-acre island was a rapidly shrinking scrap of sand about 20 feet across. We threw everything into the hatches without folding the tents or using any of the meticulous packing methods developed over the past four months. It took eight minutes to break camp, by which time the kayaks were floating in shin-deep water. With no solid land in sight and the mangroves flooded, we looked at one another and hastily agreed: "Let's finish this thing."

I only wish we had thought to don our life jackets…and void our bladders.

I fired up the GPS and pointed my bow toward "Stop Forrest!" at S 00°55'27.9"; W 04°79'26.29". With no reference points in the mat-black void for visual reckoning, we would be relying on Instrument Flight Rules for the rest of the trip.

At first the going was easy and we were thrilled to paddle through bioluminescent sea organisms that glowed blue and green as the movement of our kayaks and blades stirred them. The tiny illuminated sea creatures were so bright we could see our faces and hands as we paddled. The water was thankfully calm and the Amazon sky was covered from horizon to horizon with stars. The nights, prior to this night, the waves and conditions were too rough to let my attention stray too far from the next oncoming wave. Now, nearing the end of our journey in this calm water, the skies quickly cleared and stars shone as bright as they had on the banks of Lago Acucocha 111 days before. Stars above and below.

The mystical moment was short lived, as the incoming tide was just getting warmed up. The sand islands, noted accurately on the GPS, were now submerged sandbars. The closer we got to the Atlantic, the bigger the waves grew. In the shallows above the sunken sandbars, the waves crested and broke at four to five feet, while the deeper channels focused the energy of the tide and swell into gauntlets of 10-foot breaking waves, gauged by the length of my 18-foot sea kayak. It was the first time we had to paddle into breaking waves, so we timed our approach and occasionally had to take one in the kisser. The deeper the water got over the flooded sandbars, the less the breakers hit us, but the swells got larger.

Every glance at the brightly-lit GPS screen caused me to temporarily lose my night vision, so I chose a star and aimed my pointed bow in its direction. When that star climbed too high to follow I'd pick another one to chase. Eventually, the moon rose. The waves continued to grow bigger but now in the moonlight we could see them coming, which turned out to be far less comforting than we hoped.

Soon we came upon more fish traps, large rings of vertical stakes driven deep into the sandy bottom to form enclosures about 50 feet in diameter. The traps resembled medium-sized rodeo arenas, with more spikes arranged in chutes open to the tidal currents to herd fish into the trap. The chutes, like a long line of fence posts, disappeared into the darkness on either side. These ingenious traps were almost as dangerous to us as they were to the fish. If we knew where the chutes ended, we would have paddled around them, but for all we knew they went on for miles.

Tentatively, we approached the tall spikes, which disappeared with every approaching wave, then shot up through the troughs between the waves as if Poseidon's trident were trying to impale our fragile kayaks. The thought of coming down on one of those spikes and having it shoved up through our kayaks, and possibly our bodies, was rather sobering, and a jolt of adrenalin replaced the caffeine we were forced to forgo during our rude awakening. Once we got within a few feet of the spike fences, we chose a gap, then timed a sprint through the four- to six-foot opening between the logs. The logs were spaced much tighter as the fence neared the circular trap, eventually forming a solid wall. This meant we had to sprint up the face of a wave that was doing its damndest to push us back onto the deadly spikes. Theoretically we had just enough space to slip between the pilings, but the rising waves often lifted us sideways as well as up, and we feared they would drop us directly onto the spikes.

Herndon described identical fish traps further upriver, near Mosqueiro:

> *"There are large mud flats near the mouth of this river which are enclosed with small stakes driven in the mud close together, for the purpose of taking fish when the tide is out. A great many small fish, about the size of a herring and called mapara, are taken and salted for food."*

Beyond the fish traps the ocean grew so much larger, making us so much smaller. Miles ahead of us to the east we saw breaking waves taller than houses. Once we spotted the lights of a passing ship bobbing on the waves as it headed south in the (holy crap) ocean. As the last vague remnant of land slowly disappeared from the edge of our peripheral vision to the far right, the Amazon gave up any lingering resemblance to a river and became an all-out, pissed-off sea. Twenty-foot rollers hit us head on. We'd long since realized the grave mistake of neglecting to don our life jackets and we were in a far too precarious condition to wrestle them from their bungee straps. Now, with death grips on our paddles and real danger at hand, the reality and isolation of our situation was at the fore.

Over several hours, our world was reduced to one moving mountain after another. Celestial lights now became intermittent as clouds moved back in. Still, in scant moonlight, our vision was relative to what lay immediately ahead. There wasn't a lot of talk, with all of our energy devoted to breathing, paddling and balancing on our 18-inch-wide rounded hulls. There was nothing to say. On the horizon, we could see occasional stars dotting the infinite black space between us and Africa. Time and again, the bright stars were blocked out by dark Jericho-sized walls of water, barely restrained from crashing down on us. Each fast-approaching peak of darkness ate up more stars than the last. We climbed hard up each steep roller, hoping, hoping, hoping that we'd crest the summit before it broke. Our calories were long spent and I'd never been more tired, but these details simply had no place in our world. Any attention drawn from our efforts was just pissing in the wind.

Now and then, I'd call out a distance to the boys, roughly translating the mileage we had to travel to end this nightmare, knowing it was the only thing on our minds. As we climbed and dove and climbed and dove, we anticipated the inevitable wave that would crash down on top of our heads. There was no way we could ride out such a nightmare wallop.

Finally, with fish traps and shallow sandbars behind us, we drew within two miles of Stop Forrest. Clouds now blocked the moon, which caused the ever-bigger waves to blend in with the even darker horizon. Ahead in the void we could feel the might of the Amazon collide with the force of the Atlantic.

I stole glances at the tiny GPS screen until we were adjacent to Stop Forrest. For miles and miles to either side of us, through the infinite night, the continent of South America fell away in a 180-degree arc and the mighty river that had been our home for 111 days became one with the Atlantic. Prying one kung-fu grip off my paddle shaft, I hit the little rubber "mark" button on the GPS to record our position. We reached the mouth of the Amazon River at 3:10 a.m. on December 5th, 2012 after paddling 4,103.2 miles.

We urged on another few strokes just to make damned sure, and then I made the call. "That's it! Turn right." This time, no one questioned my navigation. I barely got the words out when Jeff hit a hard-right rudder and we aimed for a white spit of beach to the south, the nearest land in sight. The huge waves slammed and pitched us from our left side and we did our best to let the monsters roll under us. The surfing was pretty good, with the wind and waves in our favor, but we were in no mood to enjoy the cowabunga effect that threatened to spin us sideways. The closer we got to the beach, the smaller and more frequent the waves became. Finally, Jeff and Ian hit some breakers near the shore and capsized; a "first" for the expedition. I was right on them to lend assistance, but it wasn't necessary. They simply stood up in the knee-deep water.

We dragged the kayaks up onto the sand and spent the next few minutes gathering spilled gear that washed up the beach. The winning moment came when I grabbed the last bag of Cheetos and handed them to Ian, his favorite treat. At this point we didn't know whether the tide was in or out, so we dragged the kayaks several hundred yards up the beach to the highest ground available before setting camp.

I got out the satellite phone and called Barbara to let her know we had finished. Apparently, this was not news. My entire family, Jeff's family, Ian's family and a huge crowd of people on the internet had been watching our progress via the satellite transponder on my kayak's deck. When I last spoke to Barbara we were planning to sleep through the night and finish the next morning, and we didn't have the chance to call when the rising tide abruptly rushed us into action. Luckily, she checked the internet just before going to bed and alerted everyone that we were on the move. For the last six hours, our supporters watched from afar as we honed in on the Atlantic. They were all still awake and celebrating when I called. There wasn't another soul for miles around, yet we were far from alone.

We stumbled awake under the bright morning sun, well after sunrise. The beach was just as barren in daylight as it was in the dark and we found no signs of a road or civilization, save for a handful of empty grass-roofed fisherman's huts. As we reconnoitered, Sheila reported that for several hundred dollars cash, someone loosely affiliated with the hotel would meet us at the end-of-the-road town of São Caetano de Odivelas and haul us and our kayaks back to Mosqueiro. The proposed rendezvous was some 29 miles back upriver.

>"Ah, but a man's reach should exceed his grasp,
>Or what's a heaven for?"
>Robert Browning

The goddamned Atlantic Ocean. Finally. Photo by West Hansen

EPILOGUE

"I was so worn out when we arrived, that, although I had not heard from home, and knew there must be letters here for me, I would not take the trouble to go to the consul's house to seek them. Sending Mr. Potter and the Frenchman ashore to their families, I anchored in the stream, and, wrapping myself in my blanket, went sullenly to sleep."
William Herndon, April 11, 1852 – end of journey

We were exhausted, dangerously low on water and, though we'd descended the Amazon from source to sea, we still weren't finished. We were all in a serious funk, especially Ian, who is rarely prone to fits of depression, but we had no other option. Mustering the same perseverance that got us here, we began the familiar ritual of packing the kayaks when Ian discovered an unopened bottle of drinking water, a rare bright spot. Now we had a quart to share during the 29-mile slog.

The water was calmer near the shore, and in the daylight, it was clear that we could easily have skirted the fish traps that threatened us in the darkness by staying close to shore rather than aiming straight for Stop Forrest. Off to the north, we watched the fish traps being pummeled by the same large waves we battled a few hours earlier.

We plodded along unenthusiastically for an hour before spotting a 30-foot fishing boat chugging into a sheltered inlet. We caught up and asked for a ride to Caetano Odivelas, or anywhere else. The captain, a thick, well-weathered man named Paulo, was skeptical at first but his three warm-hearted crewmen convinced him to invite us aboard, after which he was the perfect host. We gave him all the Brazilian reals we had, which amounted to about 50 American dollars, though he hadn't asked for payment. After loading the kayaks on board, the crew shared fresh fish and rice out of a communal bowl and we reciprocated with the last of our M&Ms. The gracious crew returned the bag of candy after taking a small bit for themselves, to which I insisted they keep what remained. We then lay down on piles of fishing nets and fell asleep in the sun to the soothing pock-pock roar of the unmuffled diesel engine.

Our savior, Capt. Paulo. Photo by Jeff Wueste.

Ian and Jeff on our return from the Atlantic Ocean. Photo by Jeff Wueste

*The final trip for our kayaks, towed back into the mainland from the Atlantic from the Avoca.
Photo by Jeff Wueste*

After two nights of rest and food at the Hotel El Farol, in Mosquiro, we all headed back to Texas by various flights. During the layover in Panama, on a flight I shared with Ian, we read Grayson Schaffer's article in *Outside* magazine ("Fastest to the Atlantic Wins," January 2013). The story drew heavily on my email correspondence with Rocky, which he'd apparently shared with Schaffer, and noted that David and I declined to comment. Schaffer later declined to write a follow up article after I was released by National Geographic and free to speak.

As per our agreement, we provided National Geographic with hours of exciting video footage from the entire expedition; however they never produced anything more than a short blurb for their website, which effectively muffled the primary goal of our expedition. National Geographic would not commit to making, or not making, a documentary about our expedition so the task fell on me (*Peeled Faces on the Amazon: The Story of the Amazon Express*) after their two-year option period ended.

I'm pretty disappointed, to say the least, with National Geographic. Throughout my life I spent hours on end leafing through the yellow magazine, building dreams of far off places and adventure. Now that I've seen how some of the sausage is made, the blush is off the bloom.

At its core, the organization is a media outlet and makes its decisions with profit in mind and is not above influence from those with personal relationships, even at the expense of objectivity. Shortly after the expedition, Fox News acquired a majority holding, changed a good chunk of the staff members, then sold off their share to Disney, who now runs the show. As for this venerable grand dame of expedition publications making an independent study of the question of the source of the Amazon and/or other rivers, I'm not holding my breath.

Piotr's tireless lobbying resulted in multiple articles, in various U.S. and European publications, promoting his claim that the source of the Amazon is at the headwaters of the Apurímac, despite abundant evidence to the contrary. None of the reporters who interviewed Piotr made any attempt to contact me, Rocky or anyone else for a different perspective, though our names were mentioned in the articles. After Piotr published his own article in the *Explorer's Club Journal,* I was given an opportunity, as a member, to publish a similar length article, promoting a universal definition for the source of all rivers in order to compare apples to apples, rather than an arbitrary definition for each river.

Now, it's six years since Jeff, Ian and I plied those rough, dark waters where the Amazon and Atlantic compete for control. Jeff and I went on to complete the first kayak descent of the Volga River, in Russia, Europe's longest, in 2014. Ian has become a champion canoe racer and re-married. We're all still friends, though I don't hold my breath waiting for an invitation for high tea with John. To this day, I hold Piotr in high regard and respect, in spite of our differences. I would relish his charming company either in a lifeboat or at a black-tie dinner. I'm sure he's not pleased with some aspects of this story, but I'm sure we'll both survive.

Though Rocky and I aren't exactly exchanging holiday cards, I'll always be grateful for his discovery, or re-discovery, of the source of the Amazon and his decision to share it with me. It's somewhat ironic, given our differences, that we are now allies in the timeless battle between objective fact and political spin. Rocky's scholarship and fieldwork confirmed the new source and established, for the first time in history, the concept of a universal definition for a river's source. Piotr, who sits on the influential Flags and Honors Committee at the Explorers Club, helped Rocky become a Fellow member of the club, a position he well deserves. So, there we are.

I'm planning more expeditions and finally finishing the excruciating work of completing this journal of *what happened*, with a substantial number of relevant events taken out, for the sake of brevity and couth. Perhaps in time the accepted definition of a river's source will migrate towards the objective and universal, thus ending the rifts between Livingstone, Speke, Burton, Chmielinski, Contos, Herndon, Snow, Sobreviela and all of us who seek those faraway places.

Lastly, reader, if you're so inclined, the most distant source of the world's largest and longest river is still out there, waiting to be navigated, just a few short kilometers from where we, of the Amazon Express expedition, launched and if you decide to ascend thousands of feet through that thin air, into the frigid Peruvian Andes, to merely turn around and descend 4,200 miles through wonder and adventure to meet the Atlantic Ocean, then give me a call. I'd love to join you.

My advice to future explorers:

1) If you spend the money on good quality equipment, then you'll only cry once.
2) Robert Falcon Scott planned for everything to go right. Roald Amundsen planned for everything to go wrong.
3) Bravado is a quick road to failure. Recognize your limitations in order to succeed alongside them.
4) Sometimes an offer of help is not an offer to help you.
5) You will be down. Count on it and when you are down, stick to the routine.
6) A leader should take on the least desirable daily chore. I recommend trash collector and gear cleaner.
7) Pain, injuries and discomfort are inevitable. Letting them stop you is not.
8) Take a day off now and then, but never take three days off.
9) Be sure to have a large empty pee bottle in your tent each night.
10) Food is fuel. Don't be picky.
11) Coffee is life. Don't settle for bad coffee.
12) Partners should feel comfortable enough to say you're full of crap and smart enough to back it up with actual facts and reason.
13) Compliment and show appreciation publicly.
14) Sleep naked.
15) Don't eat with your bare hands.
16) Be ready to move at first light.
17) Take plenty of ear plugs.
18) Slow progress is still progress.
19) Don't take photos of people without their permission.
20) Talk and play with children along the way.
21) Each day is amazing, if only in hindsight.
22) Each morning, before the day takes over, find time to sit quietly, alone or with someone close, and enjoy that moment… then, go kick some ass.

Jason, Ian, West, Jeff and Erich in Mosquiro, Brazil with the Amazon behind us literally and figuratively. Photo by Sheila Reiter

About the Author

West Hansen is a native life-long Texan, proud father and husband who lives in central Texas. He was born in Pasadena and raised in League City, which used to be a quaint little town, where Friday night high school football was attended by the entire town. He spent his younger years on the gridiron, working construction and cooking jobs, rabbit hunting from the bed of a pickup truck and hanging out in Thrifty's grocery store parking lot with his buddies committing acts of philosophy. Escaping the comfort of his hometown, he attended college in the thriving metropolis of San Marcos, where after long last he was granted a degree in Psychology from Southwest Texas State University. During college he worked in a psychiatric facility and ran cattle on a ranch where he also lived. Following graduation, he worked in the public mental health field and is now a private social worker specializing in health care for people with a few winters under their belt.

After taking classes in whitewater kayaking in college, he cut his teeth on the flooded Guadalupe River, then moved on to ultra-marathon canoe racing in 1992 with his first Texas Water Safari canoe race. He set several course records in the Missouri River 340 and still holds the USCA C-2 record in the Texas Water Safari, where he has completed 20 of the 265-mile races, as of 2019. Citing an inability to do nothing, Hansen ran for U.S. Congress in 2018, but didn't make it out of the Primary, though he learned way too much about how the sausage is made. He's been a member of the Explorer's Club since 2012 and plans on spending the rest of his life exploring those faraway places, enjoying the company of family and good friends, seeing what athletic endeavors his body can tolerate and perhaps put out a few more books.

CPSIA information can be obtained
at www.ICGtesting.com
Printed in the USA
LVHW071953291221
707392LV00001B/1